BELIEVING

BELIEVING

THE NEUROSCIENCE OF
FANTASIES, FEARS, AND CONVICTIONS

MICHAEL McGUIRE

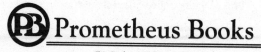

Prometheus Books

59 John Glenn Drive
Amherst, New York 14228–2119

Published 2013 by Prometheus Books

Prometheus Books recognizes the following trademark, registered trademarks, and service mark mentioned within the text: BlackBerry®; Chevrolet®; Corn Flakes®; Facebook®; Google Earth™; iPod®; Pepsi-Cola®; Prozac®; Twitter®; World Series℠.

Cover design by Grace M. Conti-Zilsberger

Inquiries should be addressed to
Prometheus Books
59 John Glenn Drive
Amherst, New York 14228–2119
VOICE: 716–691–0133
FAX: 716–691–0137
WWW.PROMETHEUSBOOKS.COM

17 16 15 14 13   5 4 3 2 1

Library of Congress Cataloging-in-Publication Data

McGuire, Michael T., 1929-
    Believing : the neuroscience of fantasies, fears, and convictions / by Michael McGuire.
        pages cm
    Includes bibliographical references and index.
    ISBN 978-1-61614-829-4 (pbk.)
    ISBN 978-1-61614-830-0 (ebook)
    1. Belief and doubt. 2. Neuropsychology. 3. Cognitive neuroscience. I. Title.

BF773.M34 2013
153.4—dc23
                                                                    2013022513

Printed in the United States of America

CONTENTS

ACKNOWLEDGMENTS

Books of this type can be unfair to their many sources. Where ideas were first read or heard are often forgotten. How discussions and advice led to textual and conceptual changes is often difficult to pinpoint. This book is no exception. For the sources and reviewers that deserve to be named here but are not, I apologize.

Many people reviewed drafts of the book or participated in discussions. These include Scott Anderson, Gary Brammer, Robert Briner, John Chandler, Toby Cronin, Damon deLaszlo, Lynn Fairbanks, Ted Harris, Steve Lorch, Roger Masters, Colleen McGuire, Katherine McGuire, Marsden McGuire, Michael Raleigh, Frank Salter, Kevin Stone, Lionel Tiger, Hap Wotila, and Arthur Yuwiler.

Four people contributed to *major* conceptual and organizational changes: John Beahrs, Jay Feierman, Beverley Slopen, and Lowell Striker. Without their contributions, the book is unlikely to have been finished and certainly not in the way it has turned out.

Beverley Slopen (the book's agent) deserves special thanks for her support and efforts in securing the book's publication. My wife, Nancy, who read and commented on each of its many drafts, deserves a special, special thanks for her efforts.

1

A CAREER TURNS UPSIDE DOWN

MRS. X

Summer in Boston is hot and humid. In my office in the hospital where I see patients, the air-conditioning drones its summer-long low hum as Mrs. X tells me once again about her persistent belief. It is a story I have heard at each of our fifteen treatment sessions. I have been practicing psychiatry for ten years, trying to treat patients with varying levels of psychic distress and feelings of inadequacy. Before that, I spent four years in medical school, one year in internship, and three years in psychiatric residency. But neither my training nor my clinical experience can help Mrs. X relinquish her strongly held belief. "I'm someone else's child," she insists.

"Let's review once more," I say. "Your birth record is signed by your mother and father. They believe you're their child. Your grandparents and your parent's friends who have known you since the day you were born believe the same thing. People say you're the spitting image of your mother. And you say your belief doesn't influence your behavior, that you love your parents, enjoy their company, spend hours with them, and still your belief is unchanged."

"That's right—Excuse me for crying."

I take a long pause, my pen floating above the pad of lined notepaper searching for something to write. I looked at her tearful face, feeling helpless. "I think it's time we accept the fact that I'm unable to provide you with the help you seek. We've talked for months. It seems clear that I don't know what to say or do to assist you in resolving your dilemma."

She continues to cry, "What about a drug?"

"I wouldn't know which drug to prescribe."

"Then what can I do?"

Again, a long pause. "We could continue to talk with the understanding that it's unlikely to help. It's not my recommendation. Another therapist is a possibility. Therapists differ in their techniques, and another technique might work. If you would like, I can give you some names. Or you might take a break for six months and then consider what to do."

"Have you talked with other therapists?" she asks.

"Yes, several. Their experiences are similar to ours. The beliefs usually persist."

After sitting there, dejected, for another few minutes, Mrs. X decided to take a break. We would meet again during winter.

Mrs. X is not the first patient I treated with beliefs that remain unchanged in the face of strong contradictory evidence. Often their beliefs and behavior are disconnected—that is, they behave as if the belief doesn't exist. Otherwise, such patients are normal and free of the everyday signs and symptoms of mental illness, such as anxiety, depression, and obsessions. Mrs. X is a typical example. She was happily married, a mother of four children, and, on matters other than her parents, was perfectly rational.

Usually I'm not easily frustrated over minor failures. Trying to fix my car and botching the job or planting a vegetable garden only to have it serve as a gourmet meal for some clandestine animal are part of living—the "fleas of life," as they are often called. But at this point, I had approached my limits. Despite nearly two decades of training and experience, I was helpless if not clueless in my ability to solve what on the surface seemed like a simple problem, altering beliefs that persist contrary to evidence in otherwise normal people. What causes such beliefs, why they persist, and what might change them had eluded me.

My wife and friends understood my frustrations, but their sympathies were no cure for my state of mind. I needed to change. I met with the chairman of the medical school department where I worked. "I need some time off to go back to basics, to study a related but simpler species. Humans are too bewildering. I want to know more about other primates. Perhaps it can lead to new treatment strategies."

We talked for an hour. He was skeptical. There were few precedents for what I was proposing. But he did grant me a leave of absence to pursue research with nonhuman primates in the hope I could satisfy my quest for answers to confounding questions of human belief and behavior.

SAINT KITTS

Three months later, I was unpacking my bags and research equipment on the island of Saint Kitts in the Eastern Caribbean. Saint Kitts and Nevis, a nearby island, are unique in that they are the present-day home of vervet monkeys (*Cercopithecus aethiops sabaeus*), a species of Old World monkeys that is prevalent throughout much of Africa. They arrived in the New World perhaps as early as 1600 CE as extra "for-sale" items on boats conducting the slave trade.

The two islands are of volcanic origin and sit atop the interface between two tectonic plates. Abundant rain feeds dense rainforests on the slopes of the volcanoes. There vervets live, breed, and thrive largely free of human interference.

The first task was to survey the island and locate good spots for observing the monkeys. It was during the third week that I encountered an elderly man carrying bananas down a mountain trail.

We talked. He was sixty-four and a native of Saint Kitts. His occupation: "I roam the hills for food that I can sell, like these bananas."

"Then you must know the monkeys pretty well." I replied.

"I do. I've lived with them all my life. Two are my pets and friends. One is thirty-two years old, but he'll not be with me long. He's ill."

"I'm sorry. Losing friends hurts. We had a cemetery for our dogs and cats on the ranch where I grew up."

"You buried your friends?" he asked.

"Yes. Why?"

"The monkeys bury their own, you know?" he said with great assurance.

"They what?" I asked.

"Yes, they bury their dead," he said again.

"How do you know?"

"You never find their dead bodies or bones. Have you?"

"Well, no, but I've only been on the island for three weeks."

"You won't find any." He was smiling now.

"Has anyone seen these burials?"

"I don't know."

"Have you?"

"No."

"Has anyone you know seen them?"

"No."

"But what about other explanations?" I asked.

"Like . . . ?"

"Well, the island is filled with animals that kill and devour dead animals. There are land crabs, carnivorous birds, ants, and probably more. Anyway, I've not seen a dead rat, mongoose, lizard, or bird. Something must be eating them."

"But there are many monkeys. They are large like dogs. There would be bones and skulls. Someone would have seen them unless they are buried."

My studies began.

During the following eighteen years, I spent three months each year studying vervets on Saint Kitts. Did I ever see a dead vervet or evidence of one's remains? No.

Do I believe they buried their dead? No.

BELIEFS

Our team conducted studies and recorded myriad findings about vervet behavior. Before dawn six days a week, we would depart for a specific location on the island, await the arrival of monkeys, and then record their behavior for five or six hours. There were moments of excitement that accompanied new discoveries, and there were moments of disappointment when the bottom fell out of "hot ideas." But for me, trying to tease out information on beliefs, hierarchy, and especially the role of the brain chemical serotonin held special fascination.

To ask it simply: Do monkeys, like humans, have beliefs? No evidence of this had been established when I arrived on Saint Kitts. But it soon appeared that they might and that their beliefs were closely tied to their behavior—that is, not disconnected from it. There were examples of beliefs that we might catalogue. The presence of humans predictably leads to their departure. Yet they are at home in the presence of land crabs, peacocks, turtles, mongooses, lizards, thousands of different bugs, and a range of island birds. They prefer group membership to living alone. Only once did I observe an animal—an elderly and graying male—living a solitary life. They have "friends" with which they spent most of their time. Certain foods are preferred and those that contain poisons are avoided.

Do they believe that humans are dangerous and many of the island's indigenous animals are not? Possibly. Do they believe that group living is desirable to living alone? Seemingly so. Do they prefer certain animals to others? Yes. Do they believe that certain foods are dangerous to their health? Also seemingly so.

The predictability of their behavior is striking. The presence of a human literally always leads to their rapid departure. Friends regularly seek close proximity to one another. The availability of specific foods leads to an interruption in what they are doing to enjoy a meal. The "disconnects" between beliefs and behavior like those of Mrs. X and others are nowhere to be found. Something is occurring in humans that is absent among their distant relatives.

HIERARCHIES

Like many species of nonhuman primates, vervets have separate male and female hierarchies. There is a dominant animal for each sex. Their behaviors differ, however. Subordinate females have close relationships with dominant females and provide them with company, grooming, and babysitting services. Challenges to a dominant female's status are rare, as is physical conflict.

Male hierarchies work another way. The struggle for who is and will be dominant occurs daily. Dominant males behave as if they believe that maintaining their status requires frequent assertion of their place. This they do by spatially displacing subordinate animals or by threatening them.

Most of the time, subordinate males respond submissively. Physical contact is likely if they don't. But at other times, they challenge dominant males with threats. The dynamic is played out daily, multiple times. Perhaps they believe that their failure to challenge a colleague may bypass an opportunity to dethrone the dominant male.

These findings aren't novel to vervets. Other species behave in similar ways. Nonetheless, their behavior is consistent with the idea that males believe that dominant status is far more desirable than subordinate status. There is no competition to remain or become subordinate.

SEROTONIN

I was struggling to make sense of these observations when a visiting biochemist alerted me about a noninvasive technique for assessing the chemical makeup of the brain via analysis of an animal's cerebral spinal fluid, the fluid that surrounds the spine. The source of this fluid is the brain, and it can be acquired without endangering an animal. A host of chemical events that occur in the brain are revealed in the composition of the fluid.

The neurochemical serotonin is a case in point. It is manufactured in the brain, where it acts to increase or decrease the frequency of certain behaviors, influences the intensity of certain emotions, and affects the interpretation of information. During these actions, it is metabolized to several byproducts, including 5-Hydroxyindolacetic acid (5-HIAA), which migrates to the cerebral spinal fluid. The greater the amount of 5-HIAA in the spinal fluid, the greater the serotonin activity in the brain.

What I'm describing may sound complex, but it isn't. In principle, it is no different than examining the contents of a neighbor's garbage can and inferring what has been eaten from what you find. If their garbage is full of artichoke leafs but there are no corncobs, an obvious inference is that they have been eating artichokes and not corn.

New techniques invite new questions. For example, does brain serotonin activity differ between dominant and subordinate males? If so, might these differences affect their beliefs and behavior? Heady questions were begging for answers. We set out to answer them.

Once every two weeks, small samples of spinal fluid were extracted from dominant and subordinate males among stable groups. The process of analyzing the fluid for 5-HIAA began. Results would be available in two or three days. Such analyses are subject to the inadvertent introduction of human error, which requires intense concentration and focus to avoid, so much so that one often forgets why the fluid is being studied.

When the results arrived, despair set in. "They can't be true," was our first response. There was no reason to believe that dominant males have twice the activity of the brain serotonin compared to subordinate males. Differences of such magnitude had not been identified previously nor had they been anticipated.

Caution prevailed. It was essential to repeat the study. The differences could be due to an error in the analysis of the fluid or unrecognized factors. Caution disappeared when the same two-to-one serotonin differences were found with the second repetition of the study. And the third, and the fourth.

Because discoveries don't occur every day, they trigger novel responses. At first we experienced doubt and confusion. I tried to make sense of the differences in ways that were consistent with what I knew before the discovery: namely, serotonin was reported to be a stable and unchanging chemical in the brain. If so, why was there an increase in serotonin in dominant or high-status males? Clearly, there was no easy explanation. Then, slowly, a warm and relaxing feeling spread over my body. A moment of self-appreciation and exhilaration followed. Then a sense of possession— "This is our discovery, our idea: dominance and high serotonin levels go hand in hand."

Other studies followed. For example, after assessing the level of 5-HIAA among subordinate males in groups, dominant males were removed from their groups. Soon the subordinate males began competing with each other to determine which among them will become dominant. After several weeks, new dominant males were in place. They had twice the brain activity of serotonin compared to the remaining subordinate males. As for the dominant males that were removed from their groups, their activity of the brain serotonin declines to that characteristic of subordinate males.

Why serotonin activity is significantly higher in dominant males—it's not higher in dominant females—remains unexplained to this day, as do the factors that lead to changes in its activity. Behavioral observations suggest an answer, however. The increase in serotonin activity positively correlates with the number of submissive displays an animal receives each day: the greater the number of such displays, the higher the serotonin activity in the animal receiving the displays. In effect, external information alters the brain's chemical activity.

Our findings were picked up by science journals and the mainstream press. The *New York Times*, *Newsweek*, and numerous science journals published feature articles. My colleagues and I received invitations to speak at universities and research institutes, and job offers followed.

In parallel with our research was that the first drug developed to elevate activity of the brain serotonin—Prozac—entered the medical marketplace. Among many people who are mildly depressed, it reduces the signs and symptoms of depression and changes negative and unpleasant beliefs to those that are more positive and pleasant. This adds up to a connection between the activity of the brain serotonin and emotional states and beliefs.

Again, new questions. Does Prozac lead people to receive more submissive displays from others and thereby elevate their status? Are subordinate vervet males with low serotonin activity depressed? Do dominant vervet males process information differently than subordinate males? If so, is serotonin activity responsible? And, of course, is serotonin activity higher among dominant human males compared to those who are subordinate?

As these events unfolded, I was tempted to do nothing else but try and solve what, by then, had grown into a long list of unanswered questions. But I had been away from home for months, and my head was filled with new ideas. The research of vervets would continue hand in hand with efforts to address questions among humans that I had first asked myself years earlier and with Mrs. X and other patients: "What are beliefs?" "Where do they come from?" "Are they a form of knowledge?" "How do they tie to evidence?" "Do they cause behavior?" "Why do people believe things for which there is no justifying evidence?" "How is the brain involved?"

2
WHERE TO START?

When I returned home and was again working at the university, my longtime friend Greg, an aspiring philosopher, phoned and suggested that we dine to "update our doings." We settled for our favorite spot. I left my office, walked across the Boston Common, then over Beacon Hill toward the harbor, and entered the Old Oyster House. As usual, it was crowded, noisy, and full of inviting fish smells. I mused to myself that, as usual, Greg would order lobster, white wine, and, for dessert, chocolate ice cream, and I would order oysters, red wine, garlic bread, and no dessert. The predictions held.

After a half-hour wait and our exposure to a rather amazing number of confidently asserted beliefs that are the trademark of bars at the end of the workday, we obtained a relatively secluded table on the second floor, where it was possible to hear each other. As dinner was finishing, our conversation took an unexpected turn.

Greg began. "I'm glad you're home. I've wanted to talk."

OK, I nodded.

"Actually I'm rather embarrassed about what I'm about to tell you. It seems silly. Please don't laugh."

I sensed his distress—he kept looking away as he spoke. "I won't laugh," I replied.

"It's this: I have two strongly held and emotionally charged beliefs that trouble me."

Again, I nodded. "And they are?"

"Liberals are altruistic and care for others. Conservatives are the opposite. These beliefs have been with me forever. They affect the way I think and vote. I can't seem to shake them. They seem to explain my behavior—do you think beliefs explain behavior?"

As the expression goes, "I was surprised beyond belief." It was not what I had expected. Greg's usual focus was metaphysics and its implications or, if not that, the irritations he experiences in trying to fix his 1937 Chevrolet.

"I'm not sure if they affect your behavior, but I'm sure that you're not alone. Half of America probably thinks the same way or the opposite. Is it really your beliefs that are troubling you?"

"No, it's not entirely the beliefs themselves. It's that I can't find supporting evidence."

For some people, the absence of evidence supporting a belief is not worrisome. But for philosophers in general, and Greg in particular, clearly it was. He had done his homework. Scholarly books had been consulted. Belief is not a new topic, and writings date back at least three thousand years. He had read much of folk psychology, which fills popular books with a wealth of definitions, views, and beliefs. On balance, he found these sources to be of limited help. Definitions lacked the precision he'd hoped for. How beliefs are identified was unclear. If and when they caused actions was ambiguous. How they interface with evidence was rarely discussed.

We dined weekly for several months, and what we ordered for dinner was more the same than not. Definitions, evidence, explanations, and opinions entered the mix of our discussions. After at least a dozen dinners, we were stumped: we hadn't found unambiguous evidence supporting or refuting his beliefs.

Our meetings were not a waste of time. We had given some structure to issues. We had made progress toward developing a usable definition of belief and devised a way of characterizing belief-evidence relationships.

BELIEFS

Beliefs. What are they?

There are, of course, definitions in dictionaries. Belief is a state or a habit of mind in which trust or confidence is placed in some person or thing. It is to accept as true, genuine, or real. It is to have a firm conviction as to the goodness, efficacy, or ability of something.[1] Its meaning in religion—faith in authority—is the topic of a later chapter.

On first pass, each of the definitions seemed acceptable. They have served so for centuries. They make sense—"common sense"—with the way people experience belief and how the word is used in everyday discussions.

Definitions were the easy part. Issues become more complex when our focus turned to the playing field of beliefs. People have beliefs about themselves, their family, their social group, their tribe, their clan, the nation, politics, what to do today and tomorrow, and the physical universe. They come in infinite shapes and sizes. They inform us of who we are and the way things happen. How best to garden, make love, fry an egg, and ask for a promotion at work are examples. They serve as benchmarks for judging ourselves and others. They insinuate themselves into the details of daily life and literally every news story and appraisal of social events. When we migrate, our beliefs travel with us. Through much of life, we try to convince others about the merits of our beliefs. When we die, we are remembered in part for what we believed.

People make hundreds of decisions each day. Beliefs are associated with literally every one of those decisions. They seem to explain which of today's chores it's wisest to finish first, what to purchase for grandmother's eightieth birthday, why tacos rather than hotdogs should be cooked for dinner, and why people get mad at their bosses. They provide a sense of direction. They answer questions to which, in their absence, there are no answers. They are the ideological foundation and operational blueprint of political systems. They are the core and indispensable bedrock of religion. They are integral to plans for building nations, exploring space, creating great art, and developing work that leads to the Nobel Prize. They are there when we engage in conflict and destruction.

Often they are inseparable from emotion. That one will marry, have a child, or believe that this year's crop will be better than usual are beliefs associated with positive anticipation and pleasure. Turn these examples around 180 degrees and people become unhappy and anxious.

Whatever their form and content, we can't and don't live without them. They are as much a part of human nature as our noses and ankles.

BELIEF *DIVIDES*

We needed a way to characterize beliefs that are supported by evidence from those that are not. We settled on the term *divide*. *A belief divide is an individual's perceived distance between a belief and his assessment of evidence related to the belief.* Divides may be narrow or wide or somewhere in between. It is the believer's assessment of *divides*, not what others perceive as *divides*, that is the focus here. *Divides* that others perceive frequently differ from those of believers. For example, there is a small *divide* separating belief and available evidence for those who are convinced of climate change, while the *divide* is wide for those who find the available evidence unconvincing. Or take the existent of God. For some believers, the Bible is irrefutable evidence of God's existence and the *divide* is narrow. For nonbelievers, the Bible, whatever it is, isn't a source of compelling evidence about the existence of God and the *divide* is wide.

Our interest then turned to how people assess *divides*. Accurate prediction is often taken as justification. For example, that water regularly freezes below a specific temperature justifies the belief that the physical state of water changes at a specific temperature.

The presence of a narrow *divide* doesn't mean that a belief is true, however. *Divides* may be narrow due to the selective interpretation of evidence or its misinterpretation. Misinterpretation was Othello's fate just as it is among many trusting lovers who are unexpectedly jilted. Selective interpretation, or its misinterpretation, appears to be the case among the 40 percent of college students who believe that some houses are haunted, the 30 percent of adults who are convinced that it is possible to influence the physical world through thought alone, and the 20 percent of adults who believe that it is possible to communicate with the dead.[2] Wide *divides* may be present when believers recognize that there is no evidence justifying a belief. But wide *divides* also can mislead: prior to Columbus's departure for the Americas, many Europeans were convinced that there was no land west of Europe.

Divides are not simply topics of academic interest. People frequently disregard or discount them because they prefer confidence in what they believe. Narrow *divides* are associated with confidence. Beliefs that people find important, which have wide *divides*, are sources of discomfort, ambiguity, and uncertainty. They are unwelcome emotionally and cognitively, sources of stress and worry, and inharmonious and providential regarding desired outcomes.

Efficient daily living requires decision, accurate prediction, and action. Closing *divides* is associated with these activities. Not only is the brain highly efficient at closure, but, in doing so, it also creates its own flavor of optimism. This was the case for many of the signatories of the Treaty of Versailles following World War I. They departed from the negotiating table with the satisfying conviction that world wars had come to an end.

OBJECTIONS AND CONSIDERATIONS

Greg and I were aware that our views might lead to objections by others. For example, might not many of the things people say they believe be described more accurately as "hunches"—strong intuitive feelings concerning a future event—"guesses," "speculations," "hypotheses," "propositions," or "conditional beliefs"? Fair enough. In everyday use, *belief* not only lacks specificity but also has multiple meanings, which are seldom clarified.[3]

That said, and with a few exceptions to be mentioned, when people say "I believe" or use a similar phrase such as "I am convinced," or "It's my view," I will assume that they believe what they say they believe until there is evidence suggesting otherwise.

There is also the matter of knowing what is believed. For the beliefs of others, there is no certain way of knowing. At times, people deceive and say they believe something they don't believe. Or it may seem that their behavior is caused by a specific belief, although it may not be: the atheist who prays in church is an example.[4] Thus, while divining the beliefs of others is only speculation, often decisions require that we do so. On the other hand, we can speak with considerable certainty about our own beliefs. We know of them because we experience them. That is, we are aware that we have beliefs. This is as close as we can come to their certainty.

Another matter concerns where beliefs come from. Probably most are acquired socially. This possibility is consistent with daily experience, which is punctuated by parents attempting to influence what their children believe, schooling, efforts of individuals and groups to politically and ideologically influence what others believe, and advertising, which usually amounts to trying to convince people that they need something they believe they don't need. There are, however, other possibilities to which I will come.

Yet another matter deals with whether or not beliefs cause action. Probably most people are convinced that they do. For example, say I believe that eating a sandwich will quench my feeling of hunger. If I prepare a sandwich, consume it, and my hunger disappears, I am likely to be convinced that my belief caused my behavior. But beliefs can be viewed in other ways, such as "an individual's representation of how the world is structured and works and how an individual's actions might result in specific outcomes."[5] This view leaves unanswered the question of whether beliefs cause action. Or beliefs can be conceived of as entities with material, energy, and spatial features that reside in the brain and may cause or bias action.[6] Which of these or other possibilities seems most plausible is left to later chapters. *For the present, I will proceed as if beliefs are associated only with action, and I will focus attention on how people experience, understand, and explain their beliefs.*

Beliefs and *divides* invite our attention. They beg for description and explanation of their nature. They invite inquiry into how evolution and culture have led to a brain that is seemingly committed to near-endless belief creation and *divide* reduction.

3
TYPES AND USES

I found my discussions with Greg stimulating and fruitful in framing and parsing the physical and metaphysical aspects of how we believe, and I assumed our dinners, with their explorations of belief, would continue. Perhaps together we would write a book. But it wasn't to happen. Greg received an anticipated and attractive invitation to teach for a year at the University of Rome. Although I was pleased for Greg, inside I was dejected. What to do next was now in my lap.

Possibilities danced about in my brain. Would an in-depth study of belief be worthwhile? How long might it take? Did I have the time and the motivation to take it on? Would it interfere with my obligations?

None of these questions was resolved when one evening my eight-year-old daughter—an aspiring artist—appeared with a hairbrush full of dried paint?

"Dad, how do I get the paint out of this?"

"That's your hairbrush isn't it? What happened?"

"I'm sorry," she replied.

"Sorry?

"Yes."

"I don't get it. I mean, how did the paint get on the brush?"

"I used it to scrape off paint on one of my paintings."

"That's all?"

"Well . . ."

"Go on, it can't hurt now."

"Well. For a while, I used it for my hair. Comet [our dog] got a spanking with it. On warm nights, it kept the window open. And before I got paint on it, I used it to dust things."

"Perhaps it also served as a paperweight and a toothbrush?"

"Once in a while as a paperweight. Dad, I'm really sorry."

Our discussion about the brush was mulling around in my head when I picked up an article dealing with octopus suckers. They facilitate movement, capture and hold prey and food, contain mechanical and chemical receptors for tasting, and signal to potential enemies.

At times, the brain does strange and unexpected things, such as connecting two seemingly unrelated pieces of information and thereby solving a problem. It was one of those moments: the hairbrush discussion and the octopus story suggested an answer to what to do next. Both brushes and suckers have multiple uses. Perhaps the same is true for beliefs. Perhaps, too, there may be multiple types of beliefs. Why not classify beliefs by their uses and types?

As it turned out, classifying was not the best of ideas. Within days, I had opened Pandora's box. There were no limits to uses and types. There were beliefs about how to raise children, organize one's life, work efficiently, stay healthy, spend holidays, write books, and grow leeks. They dealt with how the world does and doesn't work, how it should and shouldn't work, happiness and sadness, agreement and conflict, support and rejection of moral stance, suicide, when and how to act socially, and far more. Moreover, they were everywhere: books, magazine articles, cocktail parties, work, office discussions, television talk shows, newspapers, and radio programs were all dawn-to-dusk purveyors of beliefs.

I was not the first to open Pandora's box. Nearly two centuries ago, Charles MacKay wrote *Extraordinary Popular Delusions and the Madness of Crowds*,[1] which offers an informative and often-amusing history of the vulnerability of people to scams, along with their deep-rooted convictions about the validity of haunted houses, much of alchemy, and fortune-telling, and the popular follies of his day. A century and a half later, Allan Mazur's *Implausible Beliefs: In the Bible, Astrology, and UFOs*[2] confirms that little has changed except perhaps for what people believe. Mazur notes the following: Beliefs reflect their time and place in history. We accept the beliefs in which we were raised. Conversion from one belief to another is seldom about solitary truth seeking; rather, it is about conforming to the beliefs of a spouse or a friend. We coalesce around group identities and their shared beliefs especially if we are members of minorities or persecuted groups. In effect, social membership and solidarity, a believer's time in history, and early exposure emerge as more-influential factors affecting one's views than justifying evidence. Nothing written in the following pages contradicts these points.

The diversity, complexity, myriad uses, and differing views about belief promise that efforts to identify single uses or mutually exclusive types of beliefs are doomed from the start. Nonetheless, the effort is instructive if only because it illuminates the multifaceted nature of belief and its omnipresence in our lives.

What follows is a sampling of my efforts to bring some order to uses and types. The effort is far from satisfying or successful and deserves an F by usual grading standards. No compelling order turns up. Often the choice of types is arbitrary. Yet one critical finding emerges: beliefs insinuate themselves into literally every aspect of daily life.

SOCIAL MEMBERSHIP AND BELIEF

Humans are social animals. They seek each other's friendship, comfort, support, and respect. Joining groups is one way of fulfilling these desires. Adopting specific views may be an unstated yet necessary requirement for group membership.[3] In effect, group membership and belief often interweave transparently, and the desire to fulfill social needs can override the critical evaluation of the beliefs that one embraces.[4]

The following excerpts from an interview with a fifty-five-year-old ex-Marine and member of a cult capture interweaving features of belief and social membership.

While still a teenager, the interviewee had a history of failing in school and multiple trips to the local county's delinquency court. He viewed himself as "a bad and defiant boy, doomed for a life of breaking the law, interspersed by time in jail." At age seventeen, the court's judge was moments away from sentencing him to several years at youth camp. To avoid the sentence, he offered to join the United States Marines. The judge accepted the offer.

>   *Interviewee:* "It was an opportunity to be someone. I bought into the program fully. The Marines gave my life meaning. It was a

cult with its own customs and language. Suddenly I was a hero. I was part of the team that defended America. For the first time in my life, I was somebody. I believed.

Then I went to Vietnam. It was there that I became disillusioned. I could see our negative influence on the locals. We tried to turn young girls into barmaids and disrupt families. When I arrived home, it was clear that I was no hero. I no longer believed.

I started taking drugs and kept asking, 'Why am I living?' I wandered around for a while. Then I joined the XYX cult and repeated what I did in the Marines. I bought into the entire system and the idea that God would save me. I changed my lifestyle, what I thought and said, what I ate, how I dressed and approached other people. The cult became my world."

The interview highlights three signature features of belief. Belief and group membership are often conflated. Beliefs can permeate to the roots of one's being and affect habits, personality, emotion, and social behavior. *Divides* tend to narrow during the process of accepting a belief.

BEHAVIOR EXPECTATIONS AND BELIEF

There are beliefs that are inseparable from behavior expectations. Individuals, families, social groups, tribes, and clans—groups of all types—organize their affairs around convictions ranging from the mundane to the supernatural. They are the source of rules, rituals, and expected social behavior often for which there are no formal laws. How to queue in line at a movie or in a grocery store, when and how to pay homage to one's parents, standing to sing a national anthem or tribal chant, when and how to repay favors, what should and shouldn't be eaten, how to behave at funerals, and when and with whom certain topics can be discussed—all are examples. Those who don't comply invite criticism and, at times, ostracism.

BELIEFS WITHOUT CONSEQUENCES

There are beliefs that aren't associated with action and have no obvious consequences. Unless one is an astronomer and one's career hinges on it, believing that the universe is 4.45 or 13.7 billion years old doesn't affect one's sleep, love life, or self-image, and daily chores are the same with both numbers.

BELIEFS ASSOCIATED WITH ACTION

If there are beliefs that have no consequences, then there are also those that do. Their effects are not easily dismissed. With surprising frequency, families, friends, groups, and nations attribute their dissatisfactions with each other to differences in belief. Rarely does the story end there. Aftereffects often trail out for decades. The lingering regional and within-family animosities following the American Civil War and the current moral and political nastiness between the two Koreas and between the two parts of the former Czechoslovakia are examples. Or some people are convinced that killing doctors who perform abortions is serving humankind, while across town there are district attorneys who equally strongly believe that sending such individuals to prison is serving the human race. At times, groups come to believe that Armageddon is only weeks away and actively prepare for its arrival. Or who would argue that conflicting views were not significant contributing factors in the Hundred Years' War, the American and French Revolutions, World Wars I and II, the Cold War, and a host of ongoing religious-ideological-territorial conflicts?

It was beliefs in the utopian and profitable opportunities to be found in the American West that were associated with numerous prospectors, homesteaders, and religious groups venturing across the United States during the eighteenth and nineteenth centuries.[5] It was a combination of desperate hope and belief in Providence to which Sir Ernest Shackleton attributed his willingness to embark in a small, ill-equipped lifeboat on a near-impossible voyage across nine hundred miles of treacherous seas that separate Antarctica from South America. An entire shipload of stranded explorers survived because of the success of his trip.[6] It was the conviction that David Livingston was alive in Central Africa that led Henry Morton Stanley to undertake one of the most daring and glamorous manhunts in history.[7] Sir Richard Burton attributed his visits to Medina and Mecca in the 1850s while disguised as an Arab to his self-assurance about his linguistic talents.[8] It is belief in the possible that often accompanies national unification, excellent science, great music and literature, the exploration of space, humanitarian acts in response to natural disasters, and personal sacrifice. Such actions are not random. They are inseparable from well-entrenched beliefs.

TRANSIENT BELIEFS

Some beliefs are transient. They have a here-today-gone-tomorrow quality. The phrase "he changed his mind" captures what's involved. People change their minds about their wives and husbands, best friends, political parties, financial advisors, clothes, and Santa Claus. So too with what's wise to eat, which car is best, and where it's desirable to live. When change occurs, people often say that their newest belief might change in the future. That also is a belief.

Transient views often have their origins in new information and changes in preferences that accompany experience, aging, social relationships, and social status. Horses were faster than automobiles when the first automobile was manufactured. Then automobiles acquired greater horsepower and most people came to view the "iron horse" as faster. Or take flying: Two centuries ago, people proclaimed vigorously that it was impossible. Then came the Wright brothers, Charles Lindbergh, Wiley Post, DC-3s, 707s, 737s, and space shuttles. Or, for views that have lost much of their luster, consider those of Sigmund Freud and Carl Jung. During the first six decades of the twentieth century, many people embraced their explanations of the causes of mental discomfort and mental illness. Many entered treatment with the belief that their ailments would be cured. Different beliefs are in vogue today, such as that the origins of mental illness are to be found in clarifying interactions between genes, their expression, and the environment.[9]

BELIEFS WE WANT TO BELIEVE

There are convictions that people embrace because they find them pleasing. A utopian life, the Chicago Cubs will win the World Series, the universe is friendly, politicians will start telling the truth, and winning the lottery all qualify.

Excerpts from an interview with a sixty-year-old college graduate, self-employed male provide an illustration:

Author: "You said that you believe that human beings are essentially good and not evil. Can you elaborate?"

Interviewee: "There are life stories of which I am aware that cover goodness. Short of mental illness, those who have chosen evil can respond to love. In doing so, they realize that the essence of life is that good is a greater human force than evil. The discussions of good and evil in the Bible emphasize that essential point: humans have to be corrupted into evil. They can be uncorrupted."

Author: "Go on."

Interviewee: "For example, a next-door neighbor had a very difficult child. He was unable to accept boundaries or cooperate with his family. There were problems at school. Neighbors wouldn't let their children play with him because he was mean. It was deeply disconcerting to his parents. And do you know how they handled it? His mother out-loved him. Yes, she out-loved him! And it worked. He changed. He is now an adult and a fine and sensitive young man."

Author: "Are there events that might lead you to reevaluate your belief in human goodness?"

Interviewee: "Somalia and the hacking up of people who have different views is an example. They are innocent victims with alternative beliefs. They are deprived of food, which conveys the idea that one is not worth even a meal. In turn, victims often engage in evil acts to validate their worth. Or take the bankers who were largely responsible for the 2008 worldwide recession. They put thousands of people out in the street because of their greed. Their actions invalidate the lives of those affected. And how did they get so greedy? They lacked validation in their own lives and believed that money would provide it. Or take our current political system. Honesty is gone. It thrives on character assassination. Politicians are self-serving."

Author: "These examples haven't changed your mind?"

Interviewee: "Not yet. The essence of life is to be validated as a human being. This could still happen to Somalian warlords and the bankers and politicians."

Author: "Is it possible that you have this belief because it pleases you more than believing in evil?"

Interviewee: "That might be true."

PERSUASION AND BELIEF

There are views held by others that people actively work to change. Deliberate persuasion is one such means. Harriet Beecher Stowe's 1852 antislavery novel *Uncle Tom's Cabin*,[10] Peter Singer's *Animal Liberation*,[11] and Jonathan Foer's *Eating Animals*[12] are examples. They changed the views and moral commitments of many readers and served as stimuli for active proselytizing and, at times, antiestablishment action.[13]

The works of Karl Marx provide another example. Ninety years ago, they persuaded millions of people that a new type of egalitarian society was possible. For many people, the ideas served as the justification for the Russian Revolution, and thousands of committed individuals risked their lives (and thousands lost them) in their attempt to create a new social and political reality.

BELIEFS ASSOCIATED WITH STRUGGLE AND PAIN

Parents who embrace the view that their children should be punished physically for wayward behavior struggle when the moment to act arrives. Similar struggles arise when one is deceived by a close friend, by a spouse, or in a business transaction. Beliefs have been violated.

BELIEFS WITH AND WITHOUT EVIDENCE

Some people are certain their views reflect how their world works—that is, they understand reality. Often there is compelling evidence and near consensus about their views. Examples include the unforgiving action of gravity when one carelessly drops a piece of fine china onto a tile floor or that many centuries ago, there was a Roman Empire. Rarely do such views change. Other people harbor views that are at odds with all available evidence, such as that the dead can communicate with the living or that levitation is possible. Nonetheless, they believe them anyway.

But even when evidence is present, beliefs may differ. For example, say a dozen people who believe in miracles visit Lourdes seeking cures for their ailments. After praying for a cure, ten individuals believe they have been cured and are so judged by independent evaluators. Understandably, the two events are intimately tied together by those who prayed and experienced cure. The lessening or disappearance of their ailments is evidence that a miracle has occurred and God has interceded on their behalf. Their belief in miracles and the power of prayer explains the evidence. There is no *divide*. But now let's say that a concurrent scientific study demonstrates that the cures are not due to miracles. Rather, they occur because of known and predictable physiological effects of believing and praying, such as alterations in the immune and brain systems of believers.[14] These findings also qualify as evidence but of a different sort than that held by those who prayed. With the scientific findings in hand, it is tempting to dismiss the possibility of miracle cures. But to do so would likely have minimal influence on the beliefs of those who were cured. They have a belief and they have evidence.

WRONG BELIEFS WITHOUT CONSEQUENCES

There are beliefs that can be proven wrong by independent measures but are believed anyway. Individuals are often comfortable with what they believe, irrespective of contradictory evidence. For example, person A believes that route X gets him from home to work faster than route Y. It turns out that person A is wrong by seven minutes. Nonetheless, person A continues to believe that route X is faster than route Y, gets to work on time each day, and receives a promotion for his promptness. Sometimes *divides* don't matter.

WRONG BELIEFS WITH CONSEQUENCES

There are, however, wrong beliefs that do matter. The list is long.[15] They range from those about the "true" personalities and motives of friends and family members to supposedly fool-proof investment opportunities such as Ponzi schemes—the *Darwin Awards* documents new instances each year.[16] Or take natural resources: most people have been slow to relinquish their view of the Earth's infinite generosity with respect to uncontaminated water, arable soil, and clean air. The construction of Aswan Dam provides another example. When it was built in 1953, it was believed that the dam would significantly improve water distribution in the Nile Delta. Recently its designers have come to recognize that the opposite is taking place.[17] Then we have the 2003 invasion of Iraq by the United States, where the fundamental flaw was not military performance but the political beliefs and expectations that were put into place.

DISBELIEF

There is disbelief. It may deal with statements made by others regarding their personal accomplishments, the existence of UFOs, gods, higher powers, stories in the Old Testament, alleged motives of foreign powers, political ideologies, or tales of treasures buried in northern Arizona. Disbelief is another example of belief. Like belief, it too has no limits to its playing field.

AN INFORMATIVE CASE

What is to be made of organizations like SETI (Search for Extraterrestrial Intelligence)? For over fifty years, its members have worked without success to identify signs of extraterrestrial civilizations, yet their efforts remain active.[18]

There are, of course, multiple technical and strategic issues associated with such an undertaking. What areas of the universe should be scanned? At what frequencies should listening instruments be tuned? Is the available detection technology capable of identifying extraterrestrial signals, should they exist?

And what do those who work at SETI believe? It is uncertain how many hold the conviction that intelligent life is present elsewhere in the universe. But surely not all do. Yet their doubt may be far less important than it might seem. This could be the case if doubters believe that their efforts at detection are worthwhile, perhaps even commendable in their own right. There is a familiar ring to this type of belief. Many of us engage in behaviors because we believe that doing so has merit, even in the absence of justifying evidence or desired outcomes.

A SYNOPSIS

Hours, days, weeks, and years pass. Beliefs, predictions, and events come and go. Families, friends, jobs, and the physical environment change, as do the social and political worlds. Wars, droughts, earthquakes, coups, epidemics, and technological advances take place. As all of this transpires, we often wrap ourselves in convictions to which we assign private meanings. We treasure what we believe; at least this is so for many of our beliefs. Further, hundreds—perhaps even thousands—of new beliefs enter the social arena daily. Some will achieve the status of "truths," such as aliens living in New Mexico and that eating fish increases one's intelligence. Others will disappear in hours, such as the world imploding on a certain day, apricots causing cancer, and that soon people will be able to live forever. Evidence may or may not be relevant or even considered. *Divides* will range from narrow to wide. The young, the old, males, and females—all are vulnerable and, to all who believe their beliefs, their beliefs matter.

Through all of this, we sense that our beliefs are the principal navigators for daily life, a point that holds even for skeptics: they too believe in their skepticism.

SOME NUMBERS

Further efforts to identify uses for and types of beliefs might continue. For example, beliefs might be classified as past, present, or future oriented. Or they could be grouped as self-, other-, or event related. Another possible grouping is political, economic, religious, ideological, and physical-reality related. Nonetheless, the preceding examples seem sufficient to establish that beliefs are omnipresent, have multiple uses, and are associated with literally every decision we make and with every action we take. Moreover, the preceding examples address only the tip of the iceberg. For example, *Wikipedia* lists over seven hundred religious-ideological-political beliefs (e.g., Animalism, Aristotelianism, Atheism, etc.).[19] The University of Oregon Belief System Survey lists another two hundred.[20] Three centuries ago there were approximately five hundred Indian nations located in North America, each with its own nuanced version of nonhuman powers.[21] Today there are a reported 4,200 different faith groups/religions. In addition, there are numerous superstitions, taboos, and omens (Friday the thirteenth, the number 7, the warning to not walk under ladders) that many accept and use to guide their behavior.[22]

4

WHAT PSYCHOLOGISTS HAVE FOUND

According to academic etiquette, we should constrain from making pejorative comments about the research and ideas of colleagues, especially if they work in one's department or university, but that doesn't curtail the desire to do so or the practice.

Howard is a close friend. He is a professor of psychology, a "rogue psychologist," as some call him. We have a long history together. It began with high school, then college, and, several years later, working at the same university. In high school, we skipped classes together, drank beer on the sly, smoked cigarettes, and once competed for the affection of the same young lady. In college, we supported both reasonable and unreasonable causes, played instruments in the same group, and started a landscaping business. His way of settling political discussions is to remind me that he has been arrested three times to my one for protesting injustices.

When his frustration about his colleague's research passes his threshold and he is on the verge of voicing his irritations, he often drops by my office to vent and chat. It was one of those days, and he was in his usual dress, a baseball hat, suspenders, a mussed shirt, and a crooked bowtie. And, as usual, he leaned back in a chair and put his feet on my desk.

He began with a critique of the irrelevance of much research conducted by members of his department. Half an hour later, he'd exhausted the topic for the day and our conversation took a turn: "What are you up to?" he asked. I told him of my interest in belief and my plan to find out what psychologists have to say.

"Psychology? Why?" he asked.

"They've studied belief for decades," I replied.

"I'm dubious you'll find anything interesting."

"You're kidding?"

"No. I've read their stuff. It's interesting. But . . . but they're not addressing how the brain as a biological tissue generates and uses beliefs."

"Go on."

"It's like this, like what psychoanalysts do. They take what people say and then speculate about how imaginary systems in the brain such as *repression* make them say what they do. It's like trying to divine the machine code of a computer using only the information on its screen or looking at the surface of the earth and guessing what's inside. At times, what they say is ingenious. But it's still just speculation. Nothing about how the brain really works is revealed. Any breakthrough will have to address that."

"Are you telling me that there is nothing to be gained by looking at psychology?"

"No, but it will be necessary to be very precise about its limitations."

I sensed Howard had a point. But, for the moment, I wasn't prepared to bypass psychology and jump straight to the brain. For at least a half century, psychologists have studied belief, and their studies have been highly inventive in identifying ways in which the brain creates beliefs, which often defy conventional reason and how *divides* develop, persist, and go unrecognized. There is also a clear message in their work: how and what people believe is frequently riddled with information-processing quirks, logically suspect, and strongly biased in favor of perpetuating beliefs. These quirks hint at how the brain works. There would be far more to their work than Howard acknowledged.

A SAMPLING

In *How We Know What Isn't So* (1991),[1] Thomas Gilovich spells out reasons for the fallibility of human belief in everyday life. We misperceive and misinterpret random data when we assess things on the basis of superficial features or seeming similarity. We misinterpret incomplete and unrepresentative data when we address only a subset of relevant facts. We develop biased evaluations of ambiguous and inconsistent data when we disregard inconsistencies in findings. Our motivations often determine what we believe because it is personally rewarding. We accept uncritically the biasing effects of secondhand information when we believe rumors or unsubstantiated facts. We harbor exaggerated impressions of social support when we mistakenly assume that others share our views. We believe in the effectiveness of questionable interpersonal strategies.

A decade later, in *Why People Believe Weird Things*,[2] Michael Shermer offered a list of twenty-five ways in which the brain contributes to people believing weird things—that is, ways that narrow *divides*, irrespective of evidence. A sampling includes the following: Theory influences observation. Bold statements do not make claims true. Failures are rationalized to obscure errors. After-the-fact reasoning. Coincidence. Hasty generalization. Over-reliance on authority. Circular reasoning. And the need for certainty, control, and simplicity. He sums up his findings this way: "Smart people believe weird things because they are skilled at defending beliefs they arrived at for non-smart reasons."[3]

In 2006, Thomas Kida, in *Don't Believe Everything You Think*,[4] lists six basic mistakes characteristic of belief: (1) We prefer stories to statistics. (2) We seek to confirm our beliefs and reasoning. (3) We rarely appreciate the role of chance or coincidence in the interpretation of events. (4) We often misperceive our world. (5) We oversimplify. (6) We have faulty memories.

In the same year, Cordelia Fine's *A Mind of Its Own: How the Brain Distorts and Deceives* (2006)[5] provides a lively description of ways the brain biases perceptions and beliefs. The following is a sample therefrom: *The vain brain*—the brain manipulates perceptions. *The emotional brain*—emotions are not necessarily matched with the right thoughts. *The immoral brain*—the brain maintains our sense of moral superiority. *The pigheaded brain*—our brains are loyal to our beliefs. *The weak-willed brain*—the brain that fails to carry out good plans. *The bigoted brain*—the brain that doesn't acknowledge its bigotry. *The deluded brain*—the brain that allows all kinds of biases to enter thinking.

In 2011, Shermer's third book on belief appeared, *The Believing Brain: From Ghosts to Politics and Conspiracies—How We Construct Beliefs and Reinforce Them as Truths*.[6] Three themes are at the heart of his analysis of belief. The first is *patternicity*: the brain has evolved to perceive patterns even in response to random stimuli. The second is *agenticity*: the brain is strongly biased toward attributing intentional action as the cause of events. And third, we form beliefs first and then look for supporting evidence.

For beliefs dealing with ghosts, politics, religion, much of morality, and extraterrestrial life, Shermer may be largely right: beliefs often precede the search for evidence or, at times, there is no search at all. But such beliefs differ from many that occur in daily life. For example, we often acquire evidence first and only later attempt to make some sense of it. Say you note that a plant in your garden is failing to grow. On first pass, it's an observation, not a belief about a possible cause of the failure. Now suppose that the next day you observe a mound of dirt next to the plant. A belief is likely to follow: a gopher is the cause of the plant's condition.

The year 2011 also saw the publication of *Thinking Fast and Slow*[7] by Nobel Prize recipient Daniel Kahneman. For Kahneman, two systems serve as the backbone of his book and characterize the ways our brain handles information. System 1 is intuition, our fast, automatic, and unconscious system that makes instant judgments and leads to instant beliefs and action. System 2 is our far slower analytic system—think "rational" or "logical"—that reasons and often corrects judgments made by system 1. He stops short of saying that the brain is fundamentally irrational, but it is hard not to draw this conclusion from his examples.

Kahneman's two systems map much of what we experience, although as some authors have pointed out there is no reason to assume that there are only two systems.[8] We frequently make instantaneous judgments and develop beliefs based on minimal or questionable evidence. We also think things through and revise our instantaneous judgments. But as I will discuss presently, these systems are far from independent or separate. Rather, they are closely intertwined and highly susceptible to each other's influence. As to whether the brain is fundamentally irrational, that too is a topic that will be discussed.

A SURMISE

There is a very long list of other findings and publications by psychologists. Nonetheless, those cited above capture much of the essence of what they have identified. And who would disagree with much of what they have said? Daily we find our own brain, and certainly those of others, misinterpreting events, developing and harboring questionable beliefs, and narrowing *divides* without consideration of evidence.

Given their findings, an obvious implication is that we are stuck with a brain that often processes information in ways that result in beliefs that have minimal regard for supporting evidence—in effect, the brain can be irrational. Still, it is wise to view this possibility cautiously. Highly questionable beliefs may serve personal and social uses, such as assuring group memberships that have higher priority than worrying about the irrational features of such beliefs. But the most critical point is that studies by psychologists are primarily descriptive. That is, they describe the products of the brain's workings. From these descriptions, possible brain systems are inferred but not identified biologically. It is the identification of these biological systems that is critical if beliefs and *divides* are to be explained—Howard is right on this point.

From another perspective, to adopt rationality as the *benchmark* for assessing how the brain works has an obvious limitation. It shifts inquiry away from viewing the brain as it is. That is, the optimal way of viewing the brain's ways of processing information may not be as either rational or irrational any more than the workings of the spleen or the liver might be viewed as rational or irrational. They are what they are and do what they do. *Fundamental facts about the brain appear to be that it is made up of multiple systems that process information, create beliefs, and alter divides. These systems appear to have their own priorities and agendas, which often work independently of each other and outside the perceived control of their owners.* It is these systems and their workings that are of primary interest in the following pages.

5
LESSONS FROM HISTORY

With the findings of psychologists under my belt, I next chose history to see what light it might cast on belief.

Why history? Because it just made sense. Beliefs have been around since the beginning of recorded history and certainly long before. Some change; for example, new evidence and its interpretation were responsible for the gradual disappearance of certainty that the sun revolved around the earth. In other instances, beliefs such as the existence of higher powers have a remarkable durability despite their lack of evidence. Yet despite the difficulties of encapsulating its contribution to knowledge on believing, history might serve up a gold mine of examples of how beliefs are or are not tied to evidence, what alters *divides*, and the time frame of change.

It was not easy to contact one of my American history professors from college. He had retired and moved to another state. At first, I didn't recognize his voice. It had lost much of its vibrancy from earlier years. Clearly he had aged, but it soon became apparent that his memory and reasoning were as sharp as ever.

We talked several times, perhaps three hours in all. His comments ran something like this: "History lacks precision and definition. It's revised continually. Which version is to be believed? There are cultural and ideological consequences. FDR is an example. There are historians who view him as a savior in his time, the man who guided us through the Great Depression. There are others who view him as the initiator of socialism in America, an enemy to free enterprise. Or take the peopling of the Americas. Each year there is a new interpretation. Archeological facts are disputed and historians fight with each other over their interpretation—it's not pretty. And who knows what the story will be tomorrow? Or, for another example, look at the change in interpretation of America's founders. Fifty years ago, they were heroes. Today, for many, they were villains of the people, arrogant upper-class individuals who carved out a society in which they primarily would benefit."

Then he reminded me of the first book I had written, *Reconstructions in Psychoanalysis*. It questioned whether persons undergoing psychoanalysis could accurately recall their past. Accurate recall isn't simply a problem of memory. It is also about the implications of supposed unconscious systems such as *repression* that are thought to suppress memories: if some memories are suppressed, reconstructions will be incomplete. "If so," he asked, "how will you deal with personal testimonies about past events?"

I was not overjoyed with our discussions. Several days later, I telephoned him again.

"Are you suggesting that I skip history?" I asked.

"I've mixed feelings," he replied. "No doubt there are highly accurate histories that would be valuable to you. These deal with events that can be cleanly documented, such as the signing of the United States Constitution—we know who wrote and signed it. Or take the Panama Canal. There is extensive documentation about when and how it was built. But personal testimonies are another matter. The trick will be to identify those that reasonably, accurately describe events. If you plan to build your book entirely around what people tell you, it's a dangerous strategy."

Depressing as it was, I agreed with much of what he said: there are limits to much of history. Still, history is not all fiction. Events do take place and they have consequences. Moreover, a close look might be informative, provided I exercised interpretative caution. For example, beliefs change over time. Some change rapidly while others remain unchanged often over centuries. Why? When is change based on evidence, when on interpretation, when on other factors, such as emotion or technological change? Then there are the sources of beliefs. Many are acquired socially. But what about other sources? Do we sometimes make them up? Are some selected simply because they are satisfying? Do certain beliefs affect behavior while others don't? These were questions that were begging for further inquiry.

Satisfied that history might at least partially inform these questions, books started coming off my library shelf. My challenge would be to forge a path through selected findings and their interpretations in order to begin to nail down answers to lingering questions. Regarding the challenge, it needs to be noted that it's not the aim of this book to review the histories of belief, beliefs, or *divides*. Encyclopedists, historians, philosophers, and others already have done much of this work, and they have done it well. My focus here is on recurrent features of belief creation, the management of *divides*, and factors that hasten or constrain their change.

MYTHS-BELIEFS

History can be partitioned into periods during which cultures embrace packages of stories, evidence, imaginings, and rituals.[1] Viewed from afar, these packages often seem like little more than idle musings. To adopt this view is a mistake, however, as far, far more is involved. Cultural and religious myths-beliefs serve as overarching emotional-conceptual frameworks in which myths and beliefs acquire meaning and value and are associated with behavior. Tradition and hopes are socialized, given direction, and influence the political and social context of daily life, at times, down to its last detail. Irish and British cultural myths provide an illustration.

IRISH AND BRITISH CULTURAL MYTHS

At various moments during their history, the Irish have mythologized their origins in different ways. While their efforts represent attempts to establish cultural roots and uniqueness, they serve other ends. For example, during the 1600s, Irish historians and writers were struggling to rescue ancient Irish mythology from Scythian barbarity.[2] At the same time, their rival and often-despised neighbor, the British, mythologized their origins as Roman, thereby establishing a more ancient ancestry than the Irish. The Irish were not to be outdone, however. They quickly identified pre-Roman sources of Irish ancestry as a mixture of Phoenician-Egyptian heritage.[3]

Cultural myths often have little to do with historical accuracy. For the Irish during the 1600s, much of the creative energy devoted to changing their historical identity was a response to the politics of British imperialism. Their belief in a longer cultural development bolstered their cultural age, provided them with a superior moral stance, and served as a rational for challenging British colonial policy.[4]

Myths, like daily lives, are not static.[5] In recent centuries, the British have found ways of revising their myth by incorporating their heroes, such as the Duke of Wellington and his victory in the Waterloo Campaign; Lord Nelson of Trafalgar; the exploits and accomplishments of kings, queens, and statesmen; and the fighting of two world wars in less than half a century. For the Irish, their current revision involves efforts to immortalize their writers such as William Butler Years, George Bernard Shaw, and James Joyce. And at this moment, there is a project to reclaim Samuel Beckett by his homeland—this, despite the fact that Beckett turned his back on his native country for the majority of his adult life.[6]

Do the Irish and British believe their myths? No doubt many question the accuracy of the historical details as well as how they are portrayed— many *divides* are wide. But myths serve other agencies, such as uniting people who are distantly related genetically. They offer a structure for self-identity, ratcheting up national self-respect and instilling a sense of cultural uniqueness. They justify separate languages and serve as moral guides for the affairs of daily life.[7] Thus it is quite possible to covet and even believe myths and use them for multiple purposes irrespective of their historical accuracy. They can provide personal reward and pleasure that are of higher priority than a careful assessment of evidence. Under these circumstances, *divides* are easily ignored.

RUSSIAN CULTURAL MYTH

An equally informative example of creative mythologizing is found in the work of many of today's Russian historians. Prior to the fall of the USSR, they offered a qualified acceptance of the Communist revolution coupled with derogatory words about European society and values.[8] Acceptance evaporated with the implosion of the Communist enterprise, the destruction of the Berlin Wall, the reunification of a divided Germany, and the ensuing social and economic turmoil that engulfed the Russian economy.

Much as with their Irish and British counterparts, many Russian historians take as their charge the task of revising history and putting it to cultural uses. Recently they have turned to searching their distant past to identify new origins of the Russian people. A current focus is on the Mongol invasion and conquest during the thirteenth century. Under the Mongol Khans and their successors, Russia developed its own civilization and culture, at least that is a current scenario.[9] It may seem like a stretch to suggest that most living Russians would find it rewarding to embrace the belief that, twenty or so generations ago, their relatives were thirteenth-century Mongols. Yet it isn't unfeasible that in twenty years many Russians will certify such a view, find it attractive, defend it with vigor, and disregard possible *divides*.

THE AMERICAN MYTH

The American myth begins with the Puritans and independence from Great Britain. It has relatively stable features, such as the importance of its political system, individual freedom, and unique geographical location. Yet it too undergoes periodic revision through the reinterpretation of the ideas of the founding fathers and through accommodation to social and ideological change.[10] At times, change is so rapid that successive generations grow up with generation-specific versions of the myth. For example, a century ago, many Caucasian Americans believed that African Americans lacked sufficient intelligence to justify their status as equal members of society. For those who believed this way, there was no *divide*. That has largely changed. A century ago, women were considered too politically naive to vote. Again, for many, there was no *divide*. That too has changed. A century and a half ago, it was assumed that there were literally no limits to America's natural resources. A century and a half of environmental exploitation followed. This may have started to change. For many, the *divide* has widened.

The American version of mythologizing its past and engaging in myth and belief revision is no different than what the Irish, British, and Russians have been doing and do. Changes in myths and beliefs occur. Yet the process is usually slow. Moreover, believing a myth doesn't mean that one is necessarily conversant with the details of the myth. For example, while most Americans probably ascribe to the American myth, only 28 percent of adults know from which country the United States gained its independence or the number of original American colonies. The 28 percent rises to 40 percent among those who are between eighteen and twenty-two years old. And how many have read the Federalist papers?

MIDDLE AGES, SATAN, AND WITCHES

Religious myths-beliefs often differ from their cultural counterparts in that they extend across cultural boundaries. In its rapid early spread, Hinduism did just that.[11] It has been so for Christianity for two millennia and more recently for Islam.[12]

Technically, myth and belief differ in their meanings. In its usual use, *myth* refers to a traditional story of ostensible historical events that serves to unfold part of the worldview of a people or explain a practice, belief, or natural phenomenon.[13] *Belief* refers to a state or a habit of mind in which trust or confidence is placed in some person or thing.[14] Yet, in daily life, the two words often intermingle in their use.

Consider the symbolic underpinning of Satan, the devil, and hell as examples of how religious myths often merge with belief. During the High Middle Ages, the presence of Satan and the devil were most marked in the then predominantly Catholic Europe, which was composed of countries with very different cultural myths. Nonetheless, believers across much of Europe struggled with their concerns about these very lifelike symbols of evil and worked to devise ways to escape their wrath.[15]

There is little that is surprising in this history. A significant number of humans have always sensed that the world is demon haunted.[16] Demons can lurk anywhere, and they assume many forms, such as representatives of the supernatural, next-door neighbors, spouses, politicians, members of ethnic groups, and agents of the Internal Revenue Service. Symbols are attached to people, animals, events, and imaginings. In the process, they acquire a special meaning and often assume the status of belief. In effect, there is a symbolism of evil that is part of human nature, and it is as much intertwined with the broad scope of everyday events as it is with religion.[17]

The story continues. In 1692 in Salem, Massachusetts, nineteen people were hanged because a handful of local citizens and authorities viewed them as patrons of the devil. Their crimes were witchcraft and the exercise of supernatural powers for evil. Black magic, sorcery, enchantment, Satanism, and related occult arts were the tools of their trade.[18] Yet the events of 1692 were hardly novel to Massachusetts. The origins of witchcraft in Europe can be traced to pre-Christian pagan cults, Roman religion, and the writings and teachings of the Gnostics.[19] Nor did Satan or the devil rapidly exit from the Christian psyche. They remained prominent features well into the eighteenth century, and there are reports of the practice of witchcraft in small enclaves of the Basque Country during the closing years of the twentieth century.[20]

While *divides* were narrow among believers through much of this history, a more important issue for this book is this: beliefs about evil forces devoted to tempting humans to sin and that lead to the disruption of their lives raise obvious questions about the pleasure and reward of such beliefs. Where's the pleasure and where's the reward? Avoiding Satan's temptations or the devil's wrath may reduce a believer's anxiety for an hour or two. Yet short-term avoidance is hardly the optimal strategy for achieving a melodious and reparative life.

An alternative explanation is needed for the centuries-long persistence of such beliefs. Possibly it is this. Views associated with the end of the world, such as doomsday dates, life in purgatory as punishment for earthly behavior or belief, and the presence of evil forces like Satan are recurring products of the human brain. Each generation creates its own versions of such views.[21] Some versions find their way into a culture's psyche. If they are created in the absence of evidence and are partially counterfactual, which they usually are, they are exceedingly difficult to disconfirm; although, paradoxically, their partially counterfactual nature may make them easy to remember.[22] Nonetheless, over time and with the absence of evidence, they gradually disappear because of the brain's biases favoring pleasure and reward over punishment, fear, and deprivation. *Divides* gradually widen and attain recognition.[23] But again, the process can be centuries long.

NEUROMARKETING

Insights into why cultural myths and beliefs develop, vary in their attractiveness, change, and sometimes disappear are suggested by findings from the field of neuromarketing.

It works this way. A group of research subjects is presented with a prospective commercial product in several different forms. Product label, container design, or taste (if a product is consumable) might be tested. During the presentation, functional magnetic resonance imagery (fMRI) technology, which identifies the rate of blood flow in areas of the brain, is used to identify areas of increased flow in response to different forms of the prospective product. The test product that is most likely to find its way to the marketplace is the one that initiates the greatest flow in the brain's pleasure and reward areas. Such evaluations are possible because external information, such as a bottle of Pepsi-Cola, reliably activates specific brain areas. In turn, increased flow in pleasure and reward areas signals the likelihood that a product will be purchased when shopping.[24]

Of course, it's not quite so simple. There are differences among people in their brain capacities, and already-present age- and sex-related consumer preferences need to be considered. The cost of the product and the features of earlier package design may constrain the degree to which products undergo change without losing their market share—the Kellogg's Corn Flakes label is altered only slightly from year to year. Still, the principle holds: people long for pleasure and reward, and the presence of pleasure and reward predicts behavior.

Is there a fundamental difference between the neuromarketing of products for the marketplace and the creation and revision of cultural myths? The answer would seem to be no. Both aim to identify and initiate responses of pleasure and reward for those who create them and those who embrace them—who covets a cultural myth or a store-bought product in which pleasure and reward are absent? A key point, however, is this: activation of the brain areas associated with reward and pleasure appears to initiate significant *divide*-reduction. This may explain, in part, the willingness of people to adopt and defend myths irrespective of justifying evidence.

How do the preceding points add up? Myths and their beliefs insinuate themselves into the detailed fabric of daily life and social conduct, perhaps close to all of it. They are embedded in customs and daily behavior.[25] They provide a framework with which to interpret and respond to both old and new information. That is, they serve as interpreters of familiar symbols, feeling, and action and, as they do so, they alter the structure and activity of the brain.[26] At times, they seem impervious to change, as is the case among groups that are committed unwaveringly to religious beliefs or individuals who refuse to part with their distorted self-evaluations. Yet beliefs do change. The length of time they endure provides a rough measure of the extent to which they serve critical roles for the emotional-cognitive lives of individuals and groups.

None of this is to suggest that people are aware of events taking place in their brains while they are absorbing and living myths. We are no more aware of many of the brain's activities than we are of those of the pancreas. However, at times, we gain hints about the brain's workings. For example, one may experience satisfaction in viewing a cultural symbol, such as one's national flag or a picture of a favorite hero. Conversely, a flag or a picture of a hero of another culture may be dissatisfying, even irritating—pictures of Hitler and Stalin had this effect on many Western Europeans during the decades following World War II. Similar points apply to beliefs whether they're political, religious, or those of a next-door neighbor.

Details about dealing with our lack of awareness of the brain's activities remain to be discussed. But already some points seem clear. The brain's unperceived information-processing systems actively create, confirm, and disconfirm beliefs. In doing so, they may or may not take evidence into account. They narrow *divides* for beliefs that please and widen them for those that displease.

COLUMBUS

At times, religious and cultural myths constrain the interpretation but not the dissemination of new evidence. New evidence receives relatively rapid public acceptance but requires years if not centuries for full cultural and religious digestion and accommodation. Columbus's voyages to the Americas provide a striking example of how news of his travels fared when introduced into fifteenth-century Europe, which was dominated by strongly held religious myths and convictions.

Prior to 1492, the majority of Europeans had no idea of the Americas. Nonetheless, despite the absence of evidence, a few individuals believed that it would be possible to reach land by sailing west from Europe. The possibility was "in the air"—John Cabot, for example, attempted to acquire funding for a trip west from Europe before Columbus set sail. Others also entertained the possibility. Martin Behaim's globe of the world and Paolo dal Pozzo Toscanelli's world chart were compatible with land existing to the west.[27] And it turns out that the believers were correct: land would be found to the west despite the fact that for many the *divide* separating the belief that there is land to the west and supporting evidence was wide.

When the news of Columbus's voyage first arrived, many Europeans quickly accepted it as fact. There were exceptions, of course. Confronted with novel and unanticipated findings of such magnitude, it's not unexpected that there would be skeptics and naysayers. They were few in number, however, and largely without political or intellectual influence. Thus, in a matter of a few years, much of Europe had learned of Columbus's discoveries (and those of Amerigo Vespucci and others) and had accepted that there was another world that they hadn't known of previously.[28] A wide *divide* had narrowed.

Despite the acceptance of evidence, the discoveries took on a tedious cadence in altering the then dominant Christian view of humanity and the European conception of history and historical process. New World Indians were not mentioned in the scriptures. Infelicitous questions about the authenticity of the Bible followed. Subsequent voyages further complicated matters. It became clear that the people inhabiting the Americas had developed complex cultures that included gods, worship, rituals, kings, administrative hierarchies, diplomats, sophisticated astronomy, calendars, writing, beliefs, and myths.[29] These discoveries undercut the European belief that theirs was the only sophisticated civilization. Centuries passed before the New World was incorporated into a revised European view of history. A similar point applies to navigation charts. Prior to 1586, charts didn't correctly account for the fact that the earth is a sphere.[30]

Myths and beliefs change over time. Most people now accept that the earth is a sphere, not a flat plain, and that the sun, not the earth, is the center of our local universe. Nonetheless, many entrenched views—habits of the brain, so to speak—tend to disappear slowly largely because they interweave with other beliefs and are immersed in the feeling, thought, and conduct of daily life. Darwin's ideas can serve as a case in point.

DARWIN AND DISBELIEF

The history of Charles Darwin's writings—"a dangerous idea," as the philosopher Daniel Dennett once noted—confers an example of beliefs that are resistant to change, at least among many.[31]

In *On the Origin of Species* (1856), Darwin's principal book, he writes:

> For natural selection acts by either now adapting the varying parts of each being to its organic and inorganic conditions of life; or by having adapted them during long-past periods of time.[32] [p. 581]

Not only did the *Origin* turn nineteenth-century biology upside down but it also created what historian Thomas Kuhn called a "paradigm shift," that is, a theory that leads to the reinterpretation of familiar evidence and also initiates the search for new evidence.[33]

As is often the case with such shifts, responses to the *Origin* were divided. Readers separated themselves into two opposing groups: those who believed that the theory made sense and those who believed it was false and a form of heresy. Overnight people who had understood the biological world one way were presented with a dramatically different and novel view: for example, organisms evolve parts that are elegantly fit to deal with the requirement of their lifestyles, such as grazing teeth, sharp predatory claws, and hooves for rapid escape over plains. Previously unrecognized processes in nature were active in shaping animal anatomy and behavior. For many, this made sense. Suddenly there was a theory that accounted to the facts. For those opposing the theory, one of its main faults lay in the implication that natural selection had been taking place for thousands if not hundreds of thousands of years. If so, how then was one to read the story of the creation of humankind and other species found in Genesis in the Old Testament?

These concerns, however, were only part of the tidings that were to come. But first, a clarification: it is not axiomatic that strongly held beliefs are coupled with intense disbelief or the rejection of alternative beliefs. For example, a gardener may be convinced that the odor of gardenias is more pleasing than that of any other flower but still appreciate the fragrance of roses as well as acknowledge that other gardeners may believe differently. This wasn't to happen with the *Origin*, however. In one part of town, many of those who believed the theory also rejected the story of creation in the Bible. In another part of town, rejection of the *Origin* was the rule among those who took the Bible as the principal and infallible source of truth. Straddling the fence wasn't possible.

It is now a century and a half since the *Origin* arrived at the bookstores in London. Recent studies show that anywhere from 24 to 85 percent of adults in different cultures believe in evolution and its primary mechanism of change, natural selection. Iceland has the highest percentage, which is greater than 80 percent, while Turkey has the lowest with 24 percent.[34]

For the United States, the percentage is 41.[35] Moreover, for those who reject the theory, their activities are far from quiescent. In 2008, Louisiana passed an antievolution bill despite protests from state and national scientific and education organizations.[36] In 2009, no fewer than a dozen antievolution bills were active in nine states—Alabama, Florida, Iowa, Massachusetts, Mississippi, Missouri, New Mexico, Oklahoma, and Texas. In 2010, five more bills surfaced in Kentucky, Mississippi, Missouri, and South Carolina.[37] As for classrooms in United States public schools, 28 percent of all teachers are advocates of evolutionary theory, 13 percent are advocates of creationism, and 60 percent advocate neither view.[38] The cultural trial of Darwin is ongoing with no obvious presentiments despite a progressively narrowing of evolution's scientific *divide*.[39]

PSEUDOSCIENCE

And what of pseudoscience? On first pass, it is tempting to view it as little more than creative examples of the human imagination without evidence— grab bags of crackpot beliefs and activities invented by charlatans and the mentally deranged, as they might be described.

A sampling of their ideas and efforts would seem to confirm this view: Teleportation is the movement of something, often a human, through solid objects or from one place to another through paranormal means. Vampires are bloodsucking creatures that leave their burial places or coffins seeking to drink blood from the living. Giants are beings with human form but of superhuman size and strength (prominent in Siamese, Hindu, Persian, Mongol, and American myths). Ghosts are the souls of dead people that appear to the living in bodily form. Animal psi describes a form of psychic understanding between animals and human beings. Reincarnation is, for those who die, rebirth as some other person or being, such as an animal. Stigmata are spontaneous developments of bruises and wounds, usually bleeding, in places corresponding to the wounds of the crucified Christ. Perpetual-motion machines are machines that run on their own energy. Carnivorous trees are trees or other large plants capable of consuming animal tissue. Exorcism is the action of expelling evil spirits by adjuring them to abandon the person, place, or object that has come under their control. Tarot cards are said to predict the future through analyzing the relationship of the cards to each other. Levitation is the raising of the human body or any object into the air without mechanical aids and thereby defying gravity. Spiritualism is the belief that there is a continuity of life after death and that the dead can communicate with the living through the use of mediums. Astrology is the divination of the supposed influences of the stars and planets on human affairs and terrestrial events by their positions and aspects.[40]

The list is revealing with regard to both the beliefs of past practitioners and their creativity. Only phrenology—the study of the brain as a composite of various organs that localize social, moral, and intellectual qualities—in a revised and updated version has stood the test of time and found its place in today's neuroscience. But this is not to say that many forms of pseudoscience were or are simply poppycock.

Certainly charlatans and the mentally deranged have earned their place in the annals of pseudoscience. Still, a close look at the research and reasoning of the practitioners is informative. It reveals their often-serious attempts to verify their beliefs and speculations using era-prevailing standards for experimentation and the interpretation of evidence. Evidence was respected and influential in leading to new methods and explanations of their findings. That is, they tried to reduce *divides*.[41]

It is now clear that highly complex and sophisticated intellectual patterns guided much of the thinking that underpinned the fields of astrology, white magic, alchemy, even witchcraft.[42] Much of it would be dismissed today, particularly that which incorporates beliefs infused with a sturdy dose of mysticism or possibly yet-to-be-identified laws of the universe.

Still, much of the evidence associated with these beliefs is acceptable. Consider alchemy. Many of its views can be traced to events we observe daily, such as things changing under specific conditions.[43] For example, water turns to ice when the temperature drops low enough. Burned wood produces heat, disappears, and becomes ash and smoke. Seeds grow into shrubs and trees and vegetables. Lizards change color. Substances such as clay or iron ore change their nature when heated. And dead loved ones communicate in dreams.

As the historian Steven Sapin has argued convincingly: "Science is a cultural activity that is an integral element of the societies in which it is practiced, and whose basic mores and conventions it shares."[44] Pseudoscience, at least some of it, closely fits Sapin's description.

In passing, it is worth noting some of the individuals who embraced these sciences: Albertus Magnus (the only philosopher to be called "The Great" and a mentor of Thomas Aquinas), Thomas Aquinas, Arnaldus de Villanova, Pope John XXII, Cornelius Agrippa, Paracelsus, Lord Alfred Russel Wallace, and Isaac Newton. Is there a more impressive roster of intellects?

While pseudoscience was most influential during past centuries—mainly those bracketed by the Renaissance—its attraction remains. Today relatively few people believe in teleportation or alien abduction. Yet it's a good bet that a third of the world's pet owners (including the author) ascribe to some form of animal psi, perhaps a tenth to stigmata and exorcism, and there are today vast and profitable industries to satisfy those committed to managing their lives with astrology and tarot cards.

An interview with a thirty-six-year-old female teacher-artist who is a university graduate, holds a master's degree in art, and who was brought up in a family with no religious affiliation provides insight into why some pseudosciences persist:

Author: "You say you believe in astrology. Fill me in if you would."
Interviewee: "I believe the sun, moon, and rising signs tell me things about others' personalities."

Author: "For example?"

Interviewee: "I don't trust Gemini. They're two-faced. They can be nice one moment and treacherous the next. And I don't like Scorpios. They are secretive and suspicious."

Author: "You know people with these signs?"

Interviewee: "Yes."

Author: "You're confident their signs describe their personalities?"

Interviewee: "They do, nearly always."

Author: "What would you do if you met a person whom you liked who was not two-faced, and then you found out he or she was a Gemini?"

Interviewee: "I'd be very cautious, even if I liked the person. I would wait to see if the second face appeared."

Author: "How do you explain the relationship between astrological signs and personality?"

Interviewee: "There is order in the universe that we can't sense or know. But it can be realized in its signs."

Author: "I don't understand. Could you elaborate?"

Interviewee: "It's [the order in the universe] analogous to gravity. You can't see or touch gravity, yet it affects everything we do. I mean, you can see the results of gravity—stones always fall down, not up. The order works the same way. You can see the results in personality."

Author: "Where did your ideas come from?"

Interviewee: "I thought you might ask that. In high school, I went to an astrologist. She told me that she could tell me the names of my friends if I could provide their signs. I did. And she correctly named three of my friends. It was amazing, truly amazing. She knew nothing about me before I visited her. I've believed in astrology ever since."

If personal experience is evidence—and it is—the preceding interview illustrates why, for many people, the belief-evidence *divide* is far narrower than might be predicted and preempts alternative views and evidence. It's easy to reason with direct evidence just as it is hard to reason against it or without it.

Critical points emerge from the history of belief. The most critical is this: whatever their source, the brain is always involved, which is to say that beliefs—all beliefs—are context bound. They are products not of a logical or rational system but systems that put things together in ways that reflect the brain's evolved internal structure and default operations.[45] These evolved structures appear to dictate that *believing something—at times, literally anything—is a fundamental default feature of the brain and that narrowing divides facilitates belief acceptance and longevity.*

6
EVIDENCE, SOURCES, AND INTERPRETATION

Evidence is the bedrock of the justice system and of scientific research. We also often imagine that it is the foundation of our own beliefs. My investigations into psychology and history yielded few satisfying answers, and I was still in a morass of contradictions when I flew to California to visit my father. Prior to his retirement, he was the president of an insurance company, and during the summers as a teenager, I worked for his company under the supervision of a claims adjustor named Joe. The experience was an eye-opener.

Joe had gone to work as a stock boy when he was sixteen. Two years later, he was transferred to the claims department. By the time we met, he had been there thirty years and was the company's chief claims adjustor with a reputation in the industry as a person with uncanny powers for identifying false claims. That summer we traveled around Los Angeles daily to interview claimants and assess their losses. For Joe, claims work was about evidence and how it ties to events.

Joe liked to eat, so we had a lunch routine. It was Mexican food on Monday, Chinese on Tuesday, German on Wednesday, hamburgers on Thursday, and on Friday, I got my choice. At lunch, we would review our findings from the previous afternoon and that morning, make decisions about the claims, and then hit the road for more interviews. His thoroughness was amazing. Literally no detail was overlooked—details that eluded me. His motto was: "If you know how the world works, spotting false claims is a cinch."

Dad and I reminisced about those days.

"I loved working with Joe. It was an education. And he always claimed I was doing a fine job, no doubt because you were president? What's happened to him?"

Dad pushed back his gray hair and smiled. "It's a classic Joe story about evidence and interpretation. When he retired, he bought a large piece of land in the foothills of the Sierra Nevada to devote his time to animal conservation. After a couple of years, he decided to increase the bird population. First, he counted the number of birds. Then he planted sunflowers, corn, and grapes, things the birds living in the area like to eat. A year later, he counted the birds again and the number had increased significantly. He was delighted."

We paused to make some coffee.

Dad continued. "It was about this time I visited him. With great enthusiasm he told me of his success, how he had established a clear connection between increasing the food supply and the number of birds. It seemed straightforward. I was convinced.

"But during my visit, Joe's friend Dave—I don't think you ever met him—dropped by. Dave works for the state's agriculture department and is an expert on birds. Joe told him of his success. Dave lauded his efforts but was skeptical. Joe wanted to know why.

"Dave was reluctant to respond. But Joe insisted. Eventually Dave provided a list of other possible explanations. I don't recall his exact words, but it went something like this: Migration patterns might have changed. Alternative feeding sites might have been compromised. Predator reduction might have allowed more than the usual number of young birds to survive. Events on the property such as a reduction in the use of machines or pesticides might have made the property more inviting."

"And how did Joe deal with that?" I asked.

"He wasn't happy."

"And then?"

"You know Joe. He never gives up easily. The next year, he counted the birds again, planted even more vegetables, checked migration patterns and predator prevalence, and increased the use of machines. More birds appeared compared to the year before."

"And?"

"I talked with him last week. He mentioned that Dave was now partially satisfied with his explanation."

Academic careers, like justice, often rise or fall on evidence. Historians are supposed to get their facts right. Chemists have to identify the elements with which they are dealing. And so forth. But academics aren't alone when it comes to the importance of evidence and its interpretation. Everyone is involved, from cooks to taxi drivers to the police, and, of course, courts and juries spend endless hours identifying, classifying, and sifting evidence.

Except for the most obvious explanations, such as sticking a finger through the hole in the roof through which water is dripping into the kitchen, there are usually alternative ways of explaining what we take as evidence. And, if we are honest with ourselves, we rarely consider these explanations once we have a belief to which we've tied the evidence, whatever its quality.

The brain uses transparent systems for organizing and interpreting evidence. Well-known examples of the workings of these systems include intuition, inference, and various types of logic. The matter doesn't rest there, however. There are other systems, many of which are idiosyncratic and often inconsistent.

It was time for me to address evidence. Since my days with Joe, I had been living with it. Yet I hadn't asked certain critical questions. What is it? Are there different types? If there are, how are they best described? And, of course, how does evidence tie to beliefs? None of these questions would lead to quick or crystal-clear answers.

POSTMODERNISTS

As postmodernists view it, life is largely "narrative." Even if we can't find truth or meaning in any "objective reality," we can still create meaning by constructing our own narratives and telling each other stories. Explanations of events are never exact. Theories are nothing more than speculations. As to evidence—something that supposedly furnishes proof—a fundamental mistake among scientists and philosophers over the last several centuries has been their conviction that there is such a thing as objective truth.[1] Despite scientific claims to the contrary, wide and often-ignored *divides* separate their explanations from evidence.

Such postmodernist trademarks are critiques primarily of modern science and its supposed progress since the Enlightenment. Unstated givens and assumptions that scientists have used to validate scientific findings since the late eighteenth century are invalid, particularly so in the social sciences. Scientific thinking and reasoning are forms of self-deception based on the assumption that there is a clear and identifiable distinction between the self and physical reality, which postmodernists insist there isn't. Facts and concepts don't exist separately from the processes of thinking and speaking about them. They are subject to the influences the brain exerts on information. They're just stories about ourselves, our beliefs, and what people take for reality. That's as good as it gets.

It's no secret that scientists, philosophers, and most everyone else engage in discussions about how best to describe and interpret evidence and identify *divides* that separate evidence from belief and explanation. It is also no secret that no scientific fact or explanation is immune from reevaluation and reinterpretation. The Newtonian constant of gravitation, which has remained largely unchanged for centuries, is now being revisited due to findings suggesting alternative values. Gravity soon may be reformulated as a special form of entropy and information storage.[2] Recently, not only has the decades-old explanation of chemical bonds come into question, but also a change in the atomic periodic table that would assign ranges in atomic weights to certain elements is on the agenda for the future.[3]

Further, if pushed, most scientists would agree with Kant's distinction between the appearance of things that are filtered through our senses and things-in-themselves ("objective reality"), which are unknowable. In effect, there are limits to what scientists can know due to the way the brain processes information. This is a pivotal assertion on which the postmodernist thesis tries to balance. Its primary implication is not what it might seem at first, however; namely, all scientific findings are invalid if things-in-themselves are taken as benchmarks. To argue this way is tantamount to asserting that nothing can be known about "objective reality" because we don't know it completely and accurately. The critical point is this: science has developed and applies a methodology that systematically reduces the probability of connecting explanations with irrelevant evidence. The method narrows *divides* when experimental findings are repeatedly confirmed. It widens *divides* when they are not. Over time, things-in-themselves can slowly change from unknowables to partial knowables. The progressive specification of the human genetic system can be characterized this way.[4] True, there are limits to what scientists can know. Yet these limits don't invalidate scientific methodology, its findings and explanations, and the predictions its explanations make possible.[5]

While provocative, the views of postmodernists are primarily those of intellectuals and artists who believe that much of the thought and evidence that has preceded them is invalid and improvident. Their ideas are only tangential to the themes discussed here, however. Why? Because they have failed to significantly inform or alter the goings-on of daily life. That rocks fall down and not up, that certain conditions are required for seed germination, and that a fire warms a cold room have been unchanged examples of such belief-evidence relationships for centuries. Claiming that water as we normally experience and understand it is not a thing-in-itself is an interesting metaphysical assertion, but from the perspective of daily life, it is little more than that. Moreover, many explanations such as why water freezes, why wood combusts and creates heat, and why seeds germinate not only accurately predict events but also have withstood the test of time. Chance or random events can hardly explain this history, which is a point postmodernists eschew.

Still the postmodernists shouldn't be dismissed out of hand. Their claim that the brain moves relentlessly toward narrative and storytelling seems accurate. There is also a sturdy ring of truth to their view when scientists assume the mantra of "experts" when they consult with institutions and government agencies. They frequently conflate personal convictions, questionable evidence, untested beliefs, and elements of cultural myths in their efforts to explain and justify social policy and design social interventions. If evidence or beliefs are questionable, the decisions and predictions to which they lead will be inexact. Much the same may be said about the contributions of scientists to TV programs such as *Nature*, *NOVA*, and those on the History channel and the National Geographic channel. A few pieces of often-questionable evidence serve as departure points for an elaborate rendering of a this-may-be-true story with the unstated implication that what is presented may accurately depict objective reality.

DIRECT EVIDENCE

Evidence is information that can be used to justify or nullify a belief. It is not a new topic. Most likely every human being past and present has had a say about it and, at times, has struggled with what it is or might be. It is a topic of discussion as often as the weather, and it can be as slippery as walking on an icy pavement.

Multiple types of evidence exist. Three are of interest here: direct, indirect, and circumstantial.

As the term is used here, *direct evidence* is the evidence of personal experience. It ranges from simple observations—a faucet is leaking—to reviewing the findings of a complex experiment, to highly stressful situations involving the full range of emotional and cognitive responses experienced when, for example, one is a passenger on an airplane in mechanical distress.

Used this way, the definition is consistent with what is observed among young children or adults who move to a novel culture or ecological setting. Largely through trial-and-error experiences and the gradual accumulation of direct evidence, they piece together an understanding of their environment, its inhabitants, and customary and expected behavior. From the early moments of life to its final hours, the importance of direct evidence as a guide to living remains high.

Yet much of what the brain processes while experiencing an event may go unrecognized and remain hidden from awareness for years. The following incident captures this point.

Years ago, I was a member of a group searching for a lost Mayan city in the jungle of northern Guatemala. There was much discussion among group members about possible dangers posed by poisonous snakes, army ants, and jaguars.

One evening, while bathing in a stream near the camp, I sensed that a jaguar was perched on the ledge above me. I turned, and there it was. "I don't have my underpants on" was my first thought. It passed quickly. The next thought went something like this: "Jaguars are a type of cat . . . cats don't like water . . . dive into the stream and swim to camp." This I did, and at that moment, there was no *divide* between my thoughts and the action that followed. Back in camp, I related the event to members of the group.

As I told them about the jaguar, I had no sense of fear or anxiety. Nor were they present for several years when I recalled the event for others.

One evening four years later, in the comfort and protection of an apartment in Los Angeles, I opened a newly purchased book about mammals of the New World. I turned to the chapter on jaguars. It described how they attack prey from above, love water, and cover their tracks by entering a stream at one point and departing at another. As I read, I was consumed with anxiety and memories of that moment in the jungle. For at least an hour, I shook and literally couldn't utter a coherent sentence.

It is unlikely that other people would have responded to the presence of the jaguar as I did. Different responses to the same experience are as common as when two people are present at a robbery or at a rock concert. One's emotional state, cognitive focus, assessment of context, and a multitude of often-ignored factors, such as haptic sensations (body activity), are associated with the ways information is processed.[6]

Critical details of how individuals respond to and interpret their experiences, how they store them in memory, and the conditions under which they can be recalled are still to be clarified. But the jaguar incident is consistent with three points. Erroneous beliefs, such as that jaguars don't like water, are associated with alterations in *divide* width. Belief may serve as a barrier to experiencing emotion. And features of experience of which one is unaware may remain unperceived for years if not for ever.

To return to daily life and direct evidence, that a sharp knife cuts tomatoes easier and more efficiently than a dull knife can be experienced and serve as direct evidence. That automobiles don't run indefinitely and must be refueled can be experienced. Beliefs build from such evidence. Many predictably predict outcomes. People regularly fuel their cars just as they select sharp knives for cutting food. About such beliefs and evidence there is near consensus. *Divides* are narrow or nonexistent. Such beliefs are the source of common-sense wisdom, which we associate with judgment, decision, and action. Those that question the wisdom can test it for themselves. And in all probability, postmodernists and now a group that identifies itself as post-postmodernists all prefer to cut tomatoes with sharp knives or, if not that, their spouses, parents, children, and friends do so.

Much of the behavior of everyday life is guided by this common-sense wisdom. Our memory serves us well in that it is easy to recall which behaviors lead to predictable outcomes and which do not. People like to solve problems, and they usually do so as efficiently as possible. Carpenters, plumbers, gardeners, cooks, housewives, arborists, doctors, lawyers, musicians, artists, and architects all have *tool kits* composed of beliefs and procedures about how the world works in familiar situations. Daily these are associated with actions that have a high degree of certainty that they will achieve intended outcomes. Further, for many people, there is considerable redundancy in their day-to-day lives. Thus, much of the time, life moves along relatively trouble-free. Tool kits that worked yesterday will work today. If they don't, they are rearranged.

However direct, a single experience usually doesn't serve as the source of a strongly held conviction. Though there are exceptions. Feeling the pain from placing one's hand in a flame or biting into a hot chili pepper for the first time usually has both immediate and lasting influence. The housewife who washes clothes with a new soap only to find that the clothes are not clean won't use the soap again. Nonetheless, most of the time, multiple direct-evidence experiences along with information from other sources are essential for the development of a belief. This is not without consequence: multiple sources means that insights dealing with why people believe some things and not others or how *divides* develop and are altered are difficult to identify.

Direct evidence can mislead. For example, there are misperceptions such as momentarily believing that a person in a crowd is a friend. These are usually corrected by gathering further direct evidence. At other times, direct evidence is subject to multiple interpretations. This occurs when people call upon different beliefs to explain the same evidence.[7] For example, persons A, B, and C observe a plant that has failed to grow. Person A believes that the failure is due to poor plant quality. Person B explains it as a consequence of inadequate soil nutrients. And person C insists that it's due to the presence of bugs in the soil. All share the same evidence. Each offers a different explanation. Presumably the plant's failure to grow could be studied and its cause clarified. However, in most daily-life situations, such studies occur rarely, which permits conflicting beliefs to persist unmodified.

Direct evidence gathered while observing an event doesn't mean that all the details of the event are perceived. This point is beautifully illustrated by the "invisible gorilla" experiment. The experiment is recounted by Christopher Chabris and Daniel Simons in their 2010 book *The Invisible Gorilla*.[8]

The study works this way: Research subjects are shown a video of people passing basketballs back and forth. They are asked to count the number of passes. The video lasts for one minute. Partway through the video, a woman dressed in a gorilla suit appears for nine seconds, pounds her chest, and disappears. With the video over, subjects are asked several questions, one of which is, "Did you see a gorilla?" Roughly half of the subjects in the original study didn't recall seeing a gorilla. The study, along with its many variations, has been repeated multiple times, each time with similar results.

The explanation provided by the authors is that the subject's failure to recognize the gorilla is due to the "illusion of attention": "We experience far less of our visual worlds than we think we do."[9] Their interpretation is consistent with a key theme of this book: direct evidence may be incomplete or misperceived. It follows that beliefs that build from direct evidence can be wrong.

There is also an identifiable hierarchy of interpretative preferences when dealing with evidence. Preferences are frequently observed between beliefs that people develop from direct evidence and those in which the source is a third person or authority. If religious beliefs are exempted, people are far more likely to favor beliefs based on direct evidence than those of a third party.[10] We prefer to believe our own beliefs and *divides* rather than those of others. Likely this explains in part why people are resistant to changing their beliefs.

INDIRECT EVIDENCE

If direct evidence is about personal experience, indirect evidence is about information from secondary sources. Books, newspapers, TV, radio, the Internet, and gossip are examples. Consider the evening news: "Country X has demanded an apology from Country Y. . . . The New York Yankees beat Baltimore 7 to 3. . . . The Dow Jones dropped five points in late trading. . . . The president is expected to land in London in an hour." Each of these statements may accurately describe what has happened or is expected to happen. Nonetheless, their source remains indirect, not direct.

Surprisingly, we often perceive indirect evidence as crisper than direct evidence. Crispness may enhance its believability and authority. Statements made on radio, on TV, in newspapers, or by next-door neighbors often are crafted such that conflicting evidence or discordant views go unmentioned; in effect, evidence that could influence *divides* or disconfirm what is reported is left unstated or selectively interpreted. This is a form of censorship, although it is seldom viewed this way. At times, sources of indirect evidence attain the status of "truth purveyors." This was the case with Walter Cronkite during the 1960s and 1970s on the *CBS Evening News*, which took on the aura of a "context of truth."

Several points follow. The most obvious is that indirect evidence often ignores the "other half of the story." Hence its crispness. A second point is that event interpretation is influenced—often significantly—by the beliefs of interpreters. This is often evident when people tell of their personal experiences. It is equally if not more strikingly evident in the areas of politics, economics, and law: the same event can be structured, interpreted, and given meaning in multiple ways. A third point is that for those whose only source of evidence is secondary, there is no foolproof way of discerning the degree to which evidence has been structured and given meaning by others. In contrast, it's easy to select examples that are consistent with what one believes and thereby narrow *divides*.

OTHER TYPES OF EVIDENCE AND EXCEPTIONS

There are other types of evidence. Legal evidence is that which is admissible in courts of law and for which there are stringent requirements. Direct evidence is admissible and, at times, indirect evidence may be too. There is also circumstantial evidence. This consists of packages of direct and indirect evidence that tend to "prove" an event or a fact by identifying other events or circumstances that afford a basis for believing in the occurrence or nonoccurrence of an event or fact. At times, it is acceptable in courts. (It is a well-established staple of mystery novels.)

For the discerning analyst, circumstantial evidence might be viewed as little more than sophisticated conjecture. Yet it often has a very different use in daily life when it is the basis for what people believe. For example, person X is a supervisor in County Y. It is public knowledge that, during the coming year, the county will be unable to pay its employees due to a shortage of funds from taxes and other sources. The county supervisors meet and recommend firing 25 percent of the county's workforce. Supervisor X is absent from the meeting. The circumstantial conclusion is that supervisor X is not supporting the proposed firing or doesn't want to be associated with it.

There is yet another side to evidence-belief relationships: sometimes believing in advance of possessing evidence pays off. That is, believing first and finding evidence second may be a more efficient strategy than the reverse, namely, acquiring bits of evidence and trying to piece together a plausible story or belief.[11]

Presumably because parents and teachers sense that this strategy invites mistakes and wastes time, it is rarely recommend. Yet in 1871, Heinrich Schliemann believed that the ancient city of Troy could be found before he set forth in search of evidence. He discovered the evidence he sought in the poems of Homer written 2,700 years earlier. They provided reliable hints about the city's location. Homer and Schliemann were right. Troy is where Homer said it would be, a few miles north of the Dardanelles.[12] A similar story applies to lasers. They began as an idea without evidence. Searches for evidence and experiments followed. Discovery was next.[13] In a similar effort, a group of scientists recently launched a plan to locate forgotten and unpublished data, possibly residing in basements and drawers of scientists around the world, that they believe may have scientific import.[14] While most scientists might be reluctant to admit it, many have followed the belief-first, evidence-second strategy with success.

There is also a case to be made for not searching for evidence. For a few decisions, such as which new home to purchase, a detailed search is probably wise. However, the often-demanding requirements of daily life and the many decisions it requires may render extensive searches for direct evidence costly and unproductive. Hence the attractiveness of indirect evidence and its explanations, such as those found in cook books, instruction manuals, road maps, or those offered by "experts."

Then there are bits and pieces of suggestive but often-unconnected evidence that are associated with various but far from compelling beliefs with indeterminate *divides*. For example, during the coming year, those who wish to do so can experience another dozen or more new articles and books questioning Shakespeare's authorship for many of the works traditionally attributed to him by ascribing them to Francis Bacon, Edmund Spenser, or Christopher Marlowe.[15]

What implications might be drawn from the preceding discussion? One is that what constitutes direct evidence is a more complex matter than is often appreciated. We believe our experiences. No other convenient choice is available. Yet direct evidence can deceive. It can be incomplete, be misleading, and undergo alterations through time due to changes in memory.[16] Further, distinguishing between what constitutes beliefs and what is evidence is not always an easy or straightforward matter. Indeed, the issue is of enough concern to the United States National Science Foundation that it has recently initiated an effort that aims to separate evidence from belief.

Another is that, despite the absence of evidence, highly accurate predictions are often possible. Most people are not experts in how fuel burns in automobile engines, yet they have figured out ways not to run out of gas. Nor is their figuring likely to improve even if they become experts in fuel combustion. So, at times, wide *divides* are present in matters that may affect understanding, yet reducing them may not improve predictions associated with beliefs.

Yet another is that people are highly dependent and responsive to indirect evidence. Again, there is no convenient alternative—daily life requires decisions even when direct evidence is unavailable.

INFERENCE AND INTUITION

On its own, evidence doesn't explain itself. We may sense that it does when our interpretations rapidly accompany experience. Most often, however, evidence appears in bits and pieces and lacks organization. Making sense of it requires sifting, organization, and interpretation. Inference—to deduce or reason from evidence to possible causes or outcomes—is one interpretation strategy.

Our inferences are most convincing when they build on direct evidence and familiar explanations or models of how the world works. For example, person T doesn't drain his outside water pipes; a subfreezing cold spell arrives, and, following that, his outside pipes burst. The inference is that the cold weather caused the water to freeze, expand, and break the pipes.

Inferences are not free of constraints and limitations. As noted, two people may have access to the same evidence, yet they develop different inferences due to unshared beliefs about how the world works.[17] As the authors of the invisible-gorilla study put it, "Our minds are built to detect meaning in patterns, to infer causal relationships from coincidences, and that earlier events cause later ones."[18] This may be so, but it doesn't mean that any two people detect meaning or infer causal relationships in the same way.

Culture can be an influencing factor. Studies show that in explaining events, Westerners are inclined to attend to a focal object, such as why two fruit trees of the same species, size, and age bear vastly different amounts of fruit, and then reason about possible causes for the difference. In contrast, east Asians are more likely to attend to broad perceptual and conceptual fields and group objects based on family resemblance rather than category membership. Or, when North Americans try to discern how a person in a group is feeling, they concentrate primarily on the person while Japanese consider the emotions of the other people in the group.[19] Similar findings are reported for people who have experienced intense religious indoctrination. Calvinists, who stress the role of the individual, show greater attentiveness to local features compared to Catholics and Jews, whose traditions stress social togetherness.[20] With this range of interpretative influences and options, it is not surprising that when two people interpret the same evidence, their interpretations rarely match exactly.[21]

Then there is intuition: a brain information-processing system that often provides ready insight and explanation and may be associated with near-instant action. *Intuitive primacy* is a term often used to refer to this system. It has been described as "human emotions" and "gut feelings" that drive judgments and action.[22]

Intuition operates "outside" awareness. It can be automatic, quick, and often highly efficient, as in situations in which one engages in on-the-spot actions. It is at work when one is walking along a crowded street that requires the rapid integration of complex information and action to avoid contact with other pedestrians.

Intuition also asserts itself in decision making, problem solving, and belief creation, even when a more rational approach is the better choice.

For example, college students are given the following problem: "A bat and a ball cost $1.10 in total. The bat costs a dollar more than the ball. How much does the ball cost?" Approximately 50 percent of students say the ball costs ten cents. The correct answer is five cents. But for many students, "ten cents feels right," they argue. Further, it is far from clear if anyone is exempt from moments of intuitive primacy or its errors. Studies suggest that people who are intuitive and in a good mood will believe almost anything.[23] This point apparently applies even among those whose professions profess the importance of rational thinking; for example, 95 percent of college professors believe that the quality of their work is superior to that of colleagues working in the same field. Ninety-five percent can't be inference. It must be intuition.

At times, inference and intuition appear to work hand in hand to reduce *divides*. Psychologists interested in the causes of wrong beliefs have identified the process of *illusory correlation*, which leads to selectively remembering more confirming evidence compared to disconfirming evidence: in effect, it's a kind of cognitive slight-of-hand that produces a desired outcome. *Data distortions* also occur: confirming cases are created—that is, imagined—and disconfirming cases are ignored.[24] Who has not reasoned this way? Further, inference and intuition often have their own agendas. This happens when there is an abundance of evidence and two people with the same belief associate their beliefs with only a selected subset of available evidence. People with strong religious and political beliefs frequently fit this description.

Scientists studying intuition point out that, without serious conscious effort at "rational thinking," intuitive primacy doesn't self-correct.[25] Possible reasons why are suggested by comparing the belief-creation and *divide*-reducing features of intuition with those of inference. They work differently. The deductive steps of inference can be recalled or evaluated another day, and their logic can be revised, if need be. Intuition too can be recalled, evaluated, and revised, but its logic remains a mystery.

But why? Why not just acknowledge far more often than we do and that all we have in hand is a mixed bag of evidence and possible explanations with indeterminate or wide *divides*? What could be more sensible? Part of the answer has been suggested: the brain has evolved to believe, and it has inbuilt systems that narrow *divides* on their own.

INTERVIEWS AND SOURCES

Insights about the sources of evidence and belief might be gained simply by asking, "What makes you believe X?" The question is asked daily, everywhere. Yet answers are seldom fully satisfying.

An interview with a thirty-two-year-old female doctor who was brought up in a family with no religious affiliation illustrates some of the difficulties in identifying sources of belief.

Author: "You mentioned Karma. Tell me more."

Interviewee: "It's hard to put my finger on it. But in essence it's this: I believe there is a force in the universe that equalizes what people do."

Author: "For example?"

Interviewee: "If you do something bad or selfish, you will experience some form of retaliation; a payback, so to speak."

Author: "You believe that?"

Interviewee: "Yes."

Author: "It's happened to you?"

Interviewee: "When I've been selfish, I always end up paying a price."

Author: "But you must know people who are selfish who don't experience paybacks?"

Interviewee: "I do. And at times, I have wondered if Karma applies to everyone. But their time will come. Karma has its own time frame."

Author: "Do you recall when you first came to believe in Karma?"

Interviewee: "Sometime in high school. I noticed that it—I mean retaliation—happened to me and my friends."

Author: "Do you recall where the idea of Karma came from?"

Interviewee: "I've often wondered about that. But I don't. Maybe it was in the air, or maybe someone told me."

Author: "Can you recall if you ever read about it?"

Interviewee: "I don't think so. But maybe I did. I can't be sure—I read a lot."

Author: "And what about your thoughts? Say you have a mean thought but don't act on it. What happens then?"

Interviewee: "It's the same. I'll pay a price."
Author: "Can you think of anything that might change your mind about Karma?"
Interviewee: "I've tried to change my mind, to get away from it. But it keeps recurring."

As in the interview above, answers to questions about the source of beliefs are many and diverse and often lack precision. There are exceptions. When individuals are asked about why they like a close friend, they usually respond with a list of positive attributes and experiences. Or those who work at stressful jobs such as policing dangerous neighborhoods, teaching in troubled classrooms, fighting wars, or controlling air traffic can be very specific about why they believe their work is stressful. The point also applies to individuals who have been indoctrinated. People brought up in highly structured religious communities or cults can usually articulate the sources of their beliefs. Revelations also qualify. Exceptions are not rules, however. Normally, source information lacks precision.

Other factors also influence interviewing. Self-deception is one: the brain often "tricks" itself into experiencing specific thoughts and feelings by altering or blocking unconscious motives or information.[26] Self-deception is not a new topic. Much of psychoanalysis is about why and how it happens and its consequences. Or, as Sartre viewed it, we are living our lives in terms of the tales we make up about ourselves and are endowing our present moment with specious significance.[27] Both Freud and Sartre would agree that self-deception is pervasive. They would agree that self-deception is associated with *divide* reduction—this is the case for example for Jean Baptiste in Camus's *The Fall* and Captain Vere in Melville's *Billy Budd, Sailor.*[28] And they would agree that its pervasiveness raises obvious questions about the accuracy of responses when people are asked to identify the sources of their beliefs.

LANGUAGE

Then there is language, with its capacity to create evidence, contribute to beliefs, and alter *divides*.

Philosophers of language are responsible for a complex body of ideas and explanations dealing with how language relates to the brain, confers meaning and truth, and serves multiple uses.[29] While their ideas are inviting topics for discussion, here is not the place to address them. Two points only are discussed: (1) how verbal statements can create *social beliefs* and (2) the impressive way such beliefs can narrow divides.[30]

Say a new school is given the name of XYZ. The naming does not describe a state of affairs. Rather, it creates a state. Provided we all agree to call the school XYZ, it can become a "social belief." That is, we come to believe it's the school's name and we conduct school-related business accordingly—literally, there is no *divide*. Social beliefs are created daily, as when we name our pets or the constellations across the sky, or when we describe others as "jerks," "narcissists," or "airheads." It's a behavior that commences before kindergarten and continues throughout life.

Social beliefs not only make up much of the world we come to believe exists and attempt to explain, but they also end up in unexpected places with questionable influence. The *Diagnostic and Statistical Manual of Mental Disorders* (DSM), published by the American Psychiatric Association and now in its fifth edition, provides an example.[31] Many diagnostic designations—that is, names given to clusters of clinical signs and symptoms—found in earlier editions have either disappeared or changed. In addition, new designations have appeared.

It is unlikely that those individuals who are responsible for revising the DSM are simply creating names for clusters without any ties to evidence. There is extensive evidence that people suffer from aversive symptoms, atypical thoughts, and behave in socially atypical ways, and that subsets of these factors often cluster. *There is also good evidence that experienced clinicians don't view the designations as entities in the real world—that is, as things-in-themselves—but as convenient ways to reference clusters of signs and symptoms even though clusters are never exactly the same for any two individuals with the same designation.* For these clinicians, the designations are not beliefs but better described as possibly helpful yet imprecise characterizations for which *divides* separating designation and evidence vary across individuals with the same designation. But it is also the case that the designations sometimes become social beliefs, at which point *divides* narrow or disappear. This happens when treatment decisions are made only on the basis of a designation, when pro-illness groups lobby the government or private foundations for research funding to study a designation, or when lawsuits follow because of alleged wrongful treatment of a social belief.

CLIMATE CHANGE

Reports, predictions, explanations, and uncertainties about climate change provide a telling example of conflicting beliefs, the fortunes of direct, indirect, and circumstantial evidence, *divides*, and political influence—yes, all of these!

The majority of scientists working in the field of climatology believe that the world's average temperature is increasing[32] and that a primary cause is excessive CO_2 emissions from the burning of fossil fuels. There is direct evidence of increased levels of CO_2 emissions compared to measures in the past. Further, there is direct evidence of changes that have occurred in parallel with elevated CO_2, such as melting glaciers,[33] rising sea levels,[34] a reduction of ocean biodiversity,[35] and a variety of other effects.[36]

It is often assumed that science and critical thinking resolve differences between what is deemed to be true and false. But this is not necessarily so. For example, there are other—but far fewer—scientists who, while acknowledging that a temperature increase may be occurring, argue that it could be a consequence of normal cyclical changes in the earth's temperature.[37] Significant rises and declines in temperature during past periods have been detected geologically, and they have occurred during periods in which today's humans or their ancestors were few in number and fossil fuels were not burned.[38] From a geologist's perspective, geological evidence is direct evidence. The inference that temperature changes may have causes other than an increase in atmospheric CO_2 due to the burning of fossil fuel is consistent with such evidence.[39]

Clearly the beliefs and evidence supporting these divergent views are not watertight.[40] Also, for many people, there has been an erosion of trust in climate-related research.[41] This is in part due to botched environmental forecasts, such as that the polar ice cap would melt by the year 2000.[42] Further, the issue has become politicized as in instances in which climatologists have been surveyed as to their beliefs and survey results cited as proof of warming.[43] Such surveys are simply popularity polls, not science.

What does this example illustrate? Many things, of course. One is that, for many people, climate change or global warming has acquired the authority of a social belief with a narrow or nonexistent *divide*. For others, an explanation-evidence *divide* regarding the causes of temperature change is present despite serious efforts of scientists to specify the width of the *divide*. Another is that once scientific issues become politicized, explanations and evidence become blurred.

One might assume that, over time, an evidence-based understanding of our world would gradually win the day. This appears to be the case within many areas of science and engineering. On average, bridges are safer today than they were fifty years ago, airplanes are less likely to have accidents than they were in 1925, and water can be desalinated more efficiently this year compared to ten years ago. Specific organisms are now known to be the cause of plague, sleeping sickness, malaria, flu, and Chagas' disease. But when it comes to our ability to convincingly tie explanation to evidence for human relationships, politics, ideologies, and many beliefs—essentially, things where evidence is difficult to accurately identify, describe, and measure—it is unclear if there is any improvement in our ability.

7
SEEING WHAT WE BELIEVE

In 1992, a group of which I was a member created a foundation for the study and preservation of rock art. Our first act was to explore rock-art paintings in the Kimberley area of northwestern Australia.[1]

The Kimberley, as it is called, is blessed with considerably more rain than much of Australia, especially at its center. There are areas of lush vegetation, large bodies of water, and a plentitude of animals such as kangaroos, dingoes, snakes, crocodiles, and bugs. Over much of the region, the runoff from the rain has carved large gorges with overhanging ledges. Beneath these ledges are thousands of unique and highly similar paintings, which are almost always red.

The paintings are known as "Bradshaws" in honor of the German archeologist who first made them known to the public.[2] Chemical analysis dates them at various ages with the most recent at 3400 BCE. At this writing, there is no archeological, historical, or mythological evidence that might identify the artists or their culture. Their origins remain a mystery to this day.

Currently, the Kimberly is inhabited by First People (Australian Aboriginals) who have their own unique painting style and whose arrival in Australia dates back approximately fifty thousand years.

Local lore was that the First People disavowed that their ancestors had painted the Bradshaws and that the paintings were created by birds using blood from their beaks. I was skeptical.

At the end of the trip, I spent a few days in the seashore town of Cairns in eastern Australia, waiting for an airplane to start the journey home. Numerous First People lived nearby. Wandering on the local beach one afternoon, I encountered a teenager and his grandfather.

"Hello, I'm Michael from the United States."

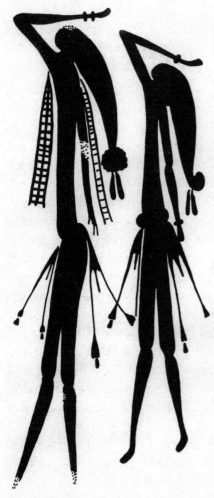

This rock art was a personal gift to the author from G. L. Walsh. This image by G. L. Walsh appears in Bradshaws: Ancient Rock Paintings of North-West Australia, *by Grahame Walsh and the Bradshaw Foundation (Carouge-Geneva, Switz.: Edition Limitée, 1994).*

"Hello, they call me B," the teenager replied. "This is my grandfather; he is also B."

"I have some pictures of paintings. Would your grandfather let me ask him some questions about them?"

The teenager spoke to his grandfather in a tongue I didn't understand. Eventually, he replied, "OK."

I showed the boy and his grandfather several pictures of First People paintings. "Are these the kinds of painting that you and your people paint?"

The boy spoke with his grandfather. Both nodded their heads indicating yes.

Then I showed them pictures of Bradshaw paintings. "Do you or your people paint these kinds of paintings?"

The boy spoke with his grandfather. "No," was the grandfather's reply.

"Does your grandfather know who painted them?"

Again a pause and a discussion in a foreign tongue. "My grandfather says they are painted by birds with blood from their beaks. They are not what he and his people paint."

"Is your grandfather certain that birds painted these pictures?" I asked.

They talked. "Yes, a bird, says my grandfather."

"Would you ask your grandfather if he sees people in these pictures [the Bradshaw pictures]?"

They examined the pictures again. Eventually the boy answered, "No. He calls them paintings of nothing. They are not people."

"Are you certain that's what he said?"

The boy repeated the question to his grandfather who nodded yes and moved his hand back and forth as if scribbling on a piece of paper. "Yes, he is certain. He says that they are not people."

Turning to the boy, I asked, "What about you? What do you see?

"People."

"Would you ask your grandfather if others his age believe as he does?"

They talked for several moments, then, "Yes."

We agreed to meet the next day to talk—I offered to bring lunch. I asked the boy to check with other elders about the Bradshaws. The next day, their responses were the same. Other elders agreed with the grandfather.

What the grandfather and other elders actually see when looking at Bradshaw paintings remains unclear. To everyone else, the paintings depict humans and reflect the human hand at work. Could what the First People see be colored by their beliefs?

A cold, hard fact of life is that beliefs are acquired from a variety of sources, such as experience, what others say, books, newspapers, radio, television, and the Internet. Another cold, hard fact is that at times we make them up, often irrespective of evidence. And yet another fact is that at times we make information or its absence fit our beliefs: *we see what we believe*.

We know that strongly held beliefs change infrequently. So, as part of the preparation for this book, I asked forty adults to identify and discuss one of their "strongly and deeply held beliefs" that had changed during the previous six months. Only two changes were reported. One person resigned from his church and joined an atheist group. The other changed his lifelong political affiliation from Republican to Democrat. None of the remaining interviewees reported changes in their deeply held political, religious, ideological, nature-of-man, or moral beliefs. Their political affiliations, convictions about God or higher powers, and views about human nature were surprisingly stable. In some instances, they had been rock solid since adolescence.

Forty adults isn't a sufficient number for a well-controlled research study. Nonetheless, their responses provide an illustration of the durability of strongly held beliefs.

The idea is not new. In 1620, Francis Bacon wrote:

> The human understanding when it has once adopted an opinion draws all things else to support and agree with it. And though there be a greater number and weight of instances to be found on the other side, yet these it either neglects and despises, or else by some distinction sets aside and rejects, in order that by this great and pernicious predetermination the authority of its former conclusion may remain inviolate.[3] [p. 112]

And why does this happen? The creation of illusory correlations—selecting confirming rather than disconfirming evidence—has been mentioned.[4] At times, confirming evidence is imagined. Or beliefs may be coveted because they are pleasurable. And so forth. One upshot of studies of belief is that the antiquated notion that knowledge arises directly from evidence has been jettisoned and superseded by a complex paradigm in which the emotional and cognitive components of belief, experience, and information from external sources interact with a bias favoring molding reality to conform to one's beliefs. In extreme instances, "beliefs tend to sustain themselves even despite the total discrediting of evidence that produced the belief."[5] An inescapable implication from these studies is that once beliefs are established, they acquire their own authority and the brain orchestrates information in ways that extend their longevity.

The tendency to see what one believes is consistent with another key theme of this book: in addition to our tool kits of familiar procedures and beliefs, the brain also harbors a ready library of beliefs in the form of models or templates that serve to order and explain information. Chapter 14 deals with these topics in detail, yet already a number of points seem clear: some models in the library are mostly innate, some are mostly learned, some are constructed from evidence, and some are imagined. Whether or not people regularly distinguish between these types is unknown, although it seems unlikely.

Decades ago, I and others conducted experiments that illustrated how beliefs color the way evidence is interpreted and how beliefs that are already present are preserved. The implications are disturbing.

COMPUTER STUDIES

Studies conducted over more than four decades ago provide an informative if not amusing example of people bringing deep-rooted beliefs coupled with confidence about their correctness to a novel experimental situation.

The studies were part of a research project designed to identify factors that lead individuals to change their beliefs. The research subjects who volunteered for the study were undergraduate students from premier American universities. They were paid to engage in conversations via a Teletype machine connected to a remotely located time-shared computer that generated replies to their typed statements. Although these studies took place during the early days of computer development (the late 1960s), it was public knowledge that local universities had designed computers that could interact with individuals at remote locations.

When subjects arrived at the laboratory, they were told only that they would be communicating via a Teletype machine for an hour, that they could discuss whatever they wished, and that they would be paid for participating in the study, irrespective of what they typed.

The software of the remotely located computer consisted of approximately one hundred rules for developing responses to the typed input of subjects. For example, if a subject typed a sentence in which the word *if* appeared, such as, "I plan to go to the beach tomorrow, *if* it doesn't rain," the computer rule was: "Disregard all words in the sentence prior to the word *if,* repeat the word *if* and the words that follow, and add the phrase *tell me more.*" For this example, the computer reply was, "*If it doesn't rain, tell me more.*" None of the computer software rules were more complex than the *if* rule. Computer replies simulated human typing speed at approximately thirty words per minute.

Described out of context, the software rules seem embarrassingly simple. Nonetheless, subjects experienced the computer's replies to what they typed as if they were communicating with another person. After an hour-long test period and anywhere from thirty to sixty subject-computer exchanges, the subjects were asked a variety of questions, including "Do you think you were communicating with a person or a computer?" Ninety percent of the subjects answered "A person." When subsequently asked, "Is it possible that you were communicating with a computer?" over 80 percent answered "No."

The research then changed in focus to try and identify the fewest number of software rules that would result in at least 50 percent of the subjects responding that they were communicating with a computer or, if not that, not with a human being. A new group of subjects participated in these studies. Over time, the software rules were degraded systematically. This resulted in computer-generated replies that omitted key words (usually verbs), made striking grammatical errors, and often made no sense. Through each degraded state, the majority of subjects continued to believe that they were communicating with another person, not a computer. When asked if they could be communicating with a computer, over half of the subjects still said "No." Some subjects volunteered explanations as to why: "Someone is trying to convince me that he is a computer," and "Computers aren't that stupid." Those subjects who believed that they were communicating with a computer commented: "People don't talk like that," "The replies didn't feel human," and "The replies were stiff."[6]

One feature of the studies wasn't part of the research design. As noted, the computer software was written to simulate average human typing speed. This it did, except at moments of computer malfunction. When this occurred, and before discontinuing operation, the computer sent the following message at a rate of 180 words per minute: "CTSS [Computer Time-Sharing System] is shutting down." Because the computer system functioned most of the time, only a subset of subjects received this message. Those who did were asked if they thought the message suggested they were communicating with a computer. None did. They offered a variety of reasons for their responses: "The person at the other Teletype must be going to lunch or the bathroom," or "That person types amazingly fast." Such responses are consistent with Michael Shermer's concepts of agenticity: that is, "the tendency to infuse patterns with meaning, intention, and agency."[7]

There are many ways to interpret the results of the studies. It was possible that another person was typing replies. There was no foolproof strategy that subjects could adopt to disconfirm this possibility. Even the "CTSS is shutting down" message could be explained as a Teletype system failure affecting both a subject and a hypothetical person at a remotely located site. Determining who or what was typing replies was further compromised by the absence of speech inflections and nonverbal gestures, which so influence face-to-face communication, TV, and radio. In short, no matter what subjects typed, they couldn't be certain they were not inter-acting with a person. Given this, the computer replies were accommodated to the belief subjects brought to the experiment: verbal exchanges take place between human beings.

In fairness to the subjects, there was no reason for them to suspect that they would be communicating with a machine. Verbal exchanges, whether face-to-face, via the telephone, over a short-wave radio, through Morse code, or via sign language all involve other people. Books and articles are primarily communications from authors to readers. Even for TV, although there is seldom direct verbal feedback from viewing audiences, it is clear that what is said is intended for other humans. From this perspective, the results of the experiment are not surprising.

Also in fairness to the subjects, these studies were conducted in the 1960s. At the time, there were few precedents for machines that commu-nicated on their own verbally or via written word. Of course, it was pos-sible to visit an amusement arcade, insert a dime into a machine, and have one's fortune told by a voice emanating from the machine. But everyone except young children knew that it wasn't the machine that was talking, but a recorded voice. Times have changed. In the decades since the 1960s, science fiction, TV dramas, movies, and industry have introduced their audiences and customers to machines that talk, make decisions, and, at times, initiate action. If the same studies were repeated today, different responses would be likely.

Still, the study's findings do not differ in principle from those of similar studies. For example, subjects can be provided with incorrect formulas that lead them to believe that spheres are 50 percent larger than they actually are. They are then asked to compare the formula-predicted volume of actual spheres with their own measurements. The discrepancy between the theo-retical and actual volumes leads to doubt, discomfort, adjustment of mea-surements, and ad hoc explanations about the discrepancy. Subjects rarely abandon their belief in the incorrect formula measurement in favor of their own measurement.[8] Formulas too can become beliefs and acquire authority.

In both the computer and the formula-sphere studies, there is a bias that favors retaining the views that are brought to novel situations. This is particularly so when prior experience has not required revision of one's beliefs: it's humans who produce words and sentences, and it's formulas that give correct answers. There is also the possibility that, even when dealing with machines, people attribute humanlike features to them and, in turn, respond as if their behavior is human generated. That is, relating to machines may be modeled after human-human interactions. Farmers frequently talk about their tractors as if they are human—"She isn't going to go to work today"—and in offices filled with computers and homes with electronic devices, it is not unusual to hear statements such as "Hurry up, damn it," "I just had you fixed," and "I didn't ask for that program." Is this different from attributing humanlike personalities to pets or supernatural entities? Or, from another perspective, does the study hint at one attraction of mathematics, that it permits an escape from anthropomorphism?

The computer study may simulate conversations in which people use only a few reply rules and yet are able to create a semblance of understanding. A familiar example is giving instructions to someone who replies with "I understand," "No, I won't do that," and so forth, only to find out later that the instructions were misunderstood. Similar events take place between teacher and student, atheist and believer, husband and wife, parent and child, and doctor and patient. One may sense that there is understanding during the interaction when, in reality, it didn't exist.

Where are the *divides* in these examples? For thirty-eight of the forty people who were interviewed about their beliefs, the *divides* for their strongly held beliefs were narrow or nonexistent. The same point holds for those who view the Bradshaw paintings as products of bleeding bird beaks. The fluctuating and unpredictable permutations of events that make up daily life had minimal effect on what they believed. For the majority of subjects in both the computer study and the wrong-formula study, *divides* also were narrow even though in both studies there was evidence that could have widened them. Interpreted this way, the findings are consistent with those who have argued that people are "bound to believe"[9] and "can't help it."[10]

I'M HEARING THINGS

The computer study is far from an isolated instance of seeing what one believes. In 1973, in a now-classic study, the psychologist David Rosenhan and his colleagues entered mental hospitals and reported to hospital staff that they were suffering from auditory hallucinations.[11] An auditory hallucination is a symptom nearly always associated with the clinical diagnosis of schizophrenia. Eight people participated in the experiment. None had a history of mental illness. All were admitted to various hospitals as inpatients. Seven were assigned the diagnosis of schizophrenia, one as manic-depressive. During their hospitalization, hospital records indicated that all were cooperative, which is unusual for people with the diagnosis of schizophrenia. It was left to the experimenters to figure out how to arrange their discharge from the hospitals. After varying periods of hospitalization, all the experimenters were discharged with the diagnosis of schizophrenia in remission.

The study is of interest for two reasons. First, similar to the computer study, the clinicians had no foolproof way of disproving the reported hallucinations. Auditory hallucinations are audible only to those who hear them, not to others, such as medical personnel who might be conducting clinical evaluations. Also similar to the computer study, medical personal apparently didn't expect that the experimenters might not fully explain their behavior. It is understandable why subjects in the computer study believed they were communicating with another person—at the time, few people had Teletype machines connected to computers. But not suspecting deception is surprising in mental-health settings: medical lore and particularly psychiatric lore stresses that patients deceive and provide incomplete and distorted histories.

Second, although in principle the several types of schizophrenia require the presence of a number of different clinical signs and symptoms to establish a diagnosis, in practice the presence of a single sign or symptom may be sufficient to initiate a diagnosis. What appears to have happened in the Rosenhan study is that the experimenters' claims of auditory hallucinations were sufficient to initiate among the clinicians a diagnosis of schizophrenia. Once the diagnosis—essentially a belief—was initiated, there was no certain way to disprove it, although the good behavior of the experimenters should have been suggestive. *Divide* narrowing likely was a contributing factor.

What is to be made of these examples other than people see what they believe?

8

RELIGION AS AN EXCEPTION TO SCIENCE—OR IS IT?

I live in the country, where the majority of land is devoted to farming or grazing. The population of the nearest town is approximately one thousand. There is regional TV, radio, and a newspaper, but much of the critical local news is communicated via the local "bush telegraph"—neighbors talk with one another. One morning, a neighbor dropped by to discuss the spread of West Nile virus in our area. As is the norm with such meetings, a wide range of topics is usually discussed. That morning was not an exception. After a time, he asked what I was doing. I told him of my work on a book about belief. A few moments in our discussion follow:

"Is your book about belief in God?" he asked.

"No, it's more general, about why and how people believe."

"Do you believe in God?" was his next question.

"No."

"That's unfortunate," he replied.

"But you do believe in God, correct?" I asked.

"Yes, all my life."

"Would you tell me why?"

"Because he watches over me."

"There is evidence for that?"

"You're a scientist. I'm not. We probably view things differently. For me, the Bible tells of real events. Then there're reports by people who have communicated with God, like Joan of Arc. I believe that Christ existed and that he is the son of God. He watches over me and influences my decisions."

"In what ways?"

"It's simple, really. I have thoughts that he puts in my head and they lead to decisions that I wouldn't make without them."

"For example?"

"Last year, I suddenly had the thought that I should pray for a good harvest. I never had that thought before. No one suggested it to me. But I prayed anyway."

"And?"

"We had an unusually good harvest."

"Let me be sure I understand what you're saying: your prayer was responsible for your good harvest? God acted on your behalf because of the prayer?"

"Yes."

"But what if your harvest had been bad?

"It would mean that I had sinned and God was not going to answer my prayer."

"But he is watching over you anyway?"

"Yes, he always does."

LINGERING CONTROVERSIES

The centuries-old controversy between science and religion is inviting to study because the proponents of differing views have possessed some of mankind's finest brains. A continuing question in these controversies is this: Do science and religion differ in the ways they develop beliefs and interpret evidence?

At first glance, this might seem to be a non-question. There are self-evident differences. Religion is based on belief in authority while science is based on the belief that evidence meets specific requirements for interpretation or that evidence can be found. But it's not that straightforward.

A close look at the controversy reveals the human tendency to create artificial categories and then assume that they are separate and unique. I was no exception. I had long harbored the belief that religion and science are distinct and separate endeavors. The twain doesn't meet. But once I began examining the details of the controversy, my categories fell apart.

EVIDENCE AGAIN

Evidence is important. At times, it confirms what people believe. At other times, it alters beliefs much as it did following the discoveries of the Americas and Christopher Columbus's voyages. At still other times, it is the goal of scientific experiments, as when chemists seek to identify the structure of a molecule. In short, some of what people believe and don't believe hinges on evidence. Some also hinges on their perception of *divides* separating beliefs and evidence. And some beliefs can't be disproved. Nowhere are these points more relevant than in the now centuries-old discourse between religion and science.

Is there a god or a higher power? No and yes. "No" for atheists. They reject the possibility because they believe there is an absence of justifying evidence. Also, perhaps, the idea may seem implausible. "Yes" for the majority of the world's adults whose confidence has its source in "*that state of mind by which it assents to propositions, not by reason of their intrinsic evidence, but because of authority.*"[1] Authority alone may be both proof and explanation. If it is accepted, there is seldom a *divide*, or at least it is very narrow. Perhaps, too, there are spontaneously created beliefs about gods and evil forces. Studies suggest that such beliefs are an inbuilt feature of human nature—that is, beliefs are products of the brain on its own.[2] If so, religion may serve to embellish and manage them.

DOMAIN OVERLAP OR SEPARATENESS?

It might be reasoned that acceptance of religious authority would settle matters dealing with evidence and belief as they apply to religion and science. They simply are not the same. Or, as Stephen J. Gould puts it, science and theology are "nonoverlapping magisteria." Gould defines magisterium as "a domain where one form of teaching holds the appropriate tools for meaningful discourse and resolution."[3] His view would separate the discourse of science from that of religion with science addressing the empirical realm and theology addressing the realm of meaning.

There are many beliefs that are consistent with Gould's view. For example, the evangelist Billy Graham asserts that "it is impossible for us who were created for eternity ever to find anything in this world to satisfy our souls."[4] Said a bit differently, thinking mystically differs from the usual rules of evidence and logic and for which options differ from those we experience on earth in daily life.[5] The two domains thus would appear to have their own ways of selecting, valuing, and interpreting information. Yet the view of nonoverlapping magisteria is not without ambiguities and critics, some of whom detect far less separation.[6]

If science's current view of evidence is taken as the benchmark, it is easy enough to agree with Gould. There is no scientifically acceptable evidence for or against God, a higher power, or an afterlife[7]—in effect, there is nothing to report one way or another. It's worth adding that no potentially informative experiments are on the drawing board. Viewed this way, there is an immediate and clear answer to the title of this chapter: yes, religion is an exception to science.

But an exception to what about science: to its current evidence requirements, to scientific method, to its explanation and reasoning, to its knowledge, or possibly to something else? Answers to these questions require a closer look at the methods and practices of science and religion, particularly as they bear on authority and possible areas of conceptual and methodological overlap. Nonoverlapping magisteria may be ripe for revision.

SCIENTIFIC METHOD AND A BIT OF HISTORY

A convenient place to begin the inquiry is by noting that the modern scientific method is barely four and a half centuries old. The 1550 publication of Copernicus's description of how the Earth and other planets in the solar system revolve around the Sun is usually taken as its inception date.[8] Why the modern method took so long to develop is an interesting question in its own right. Clearly there were many very competent scientists as far back as Babylonian times (circa 1750 BCE) and no doubt well before.[9] Part of the answer is found in the modus operandi of scientists prior to 1550. They focused primarily on *proving* their ideas. This they did with the information they had, often with ingenious reasoning, but without a well-articulated or generally accepted methodology to systematically assess their explanations and evidence. This was the case for example among many of the pseudoscientists.

Then, during the sixteenth century, the makings of a critical methodological shift began to emerge: science would gradually attempt to restructure its methodology and research strategy to that of *disproving* ideas and hypotheses rather than *proving* them.

On first pass, a change in a single word—*proving* to *disproving*—seems trivial. There were clear implications however; the methods and evaluations of empirical research were about to undergo a significant alteration, albeit slowly. The upshot of the shift was that empirical science adopted rules and procedures that ratcheted up the precision requirements for evidence and standardized research methods while simultaneously introducing a healthy dose of skepticism about science and its explanations. Evidence had to be accurately described and measured. Interpretations had to specify as precisely as possible how they explained evidence. Repeated high-quality experiments that consistently failed to *disprove* hypotheses were essential for scientists to recognize and alter *divides*.[10]

Although it wasn't fully grasped at the time, the emerging methodology identified a fundamental limitation of scientific inquiry: *adopting the disproving strategy meant that nothing could ever be viewed as proved for certain.* This limitation had unexpected consequences. Science and scientists acquired a new level of authority among the interested public. Their evidence, methods, and interpretations would gradually become available for assessment among supporters and critics alike.

A paradigm shift doesn't assure that it is universally adopted or even universally optimal. To jump ahead to the twentieth and twenty-first centuries, optimality was a central issue when scientists, philosophers, historians, and postmodernists actively critiqued scientific method and reasoning and came to view them in very different ways. Within science, the most influential writings came from the pen of physicist-historian Thomas Kuhn.[11] Kuhn argued that the *disprove* methodology not only was overstated and idealistic but also that it was often disregarded by scientists without seriously hampering their research or findings. Other scientists have voiced a similar view.[12]

For what is often referred to as the "hard sciences"—chemistry and physics—the disproving strategy may be largely workable and often optimal. But once inquiries venture to other areas of scientific inquiry, disproving hypotheses may be no more informative than efforts to prove them. For the disciplines of sociology, anthropology, psychology, and psychiatry, critical experiments often can't be conducted. There are multiple reasons why. For example, among people who experience auditory hallucinations, should evidence gathering focus on the content of their hallucinations, their frequency, their association with behavior, some combination of these, or something else? Or, if reports of hallucinations could be collected, can they be measured reliably? Another reason is that the past can't be repeated exactly. This means that much of archeology and history is a combination of bits of evidence and sophisticated inference.

The postmodernists were aware of these limitations. Thus it is hard to contest their view that much of the research scientists claim is scientific is influenced by factors over which scientific methodology has minimal or no control. Still, at times, and despite multiple limitations, research efforts do lead to interpretations that are compelling and justifiable. For example, history, archeology, and empirical science often work together and use their very different methods and knowledge to construct explanations of past events.[13] The progressive updating of the history of Stonehenge, the implosion of the Maya civilization, and the several routes taken by our ancestors during their exodus from Africa are current examples.[14] Strict adherence to the disprove-only methodology would seriously constrain such investigations.

There are other sides to this discussion. Experimental replication may be the ideal research strategy for modern science, but the findings that emerge from its use are not infallible. For example, recently, seemingly well-established and repeatedly confirmed research findings have started to look questionable, such as the effects of drugs in curing psychosis, the health benefits of vitamin E, the durability of cardiac stents, and the previously mentioned ongoing revisions dealing with chemical bonds, atomic weights, and the constant of gravity.[15] Further, how science uses findings and methods may merge with issues often considered outside its province. The scientific knowledge and methods applicable to performing abortions are well documented and understood, but the act of aborting a fetus is subject to moral, political, and religious interpretations and influence.

In short, all is not straightforward and tidy with regard to scientific method and explanation. True, for some areas of science, disproving hypotheses and studies is possible. But such options encompass only a small part of scientific effort. Thus, to maintain that science is "a domain where one form of teaching holds the appropriate tools for meaningful discourse and resolution" is to ignore the full range of scientific inquiry.[16]

ANOTHER PERSPECTIVE

There is yet another perspective that informs these issues. Scientific methodology can be viewed as a system created to outwit the brain's explanation and evidence-interpretation biases such as *divide* reduction, illusory correlations, and data distortions. There are limits to what method alone can accomplish, however. While scientists may strive to adhere to their methods, they are much like everyone else when it comes to belief creation and *divide* reduction.

For example, there are mathematicians who subscribe to a Platonic view of the universe in which theorems are true statements about timeless entities that exist independent of the human brain—things-in-themselves, as it were. Other mathematicians believe that mathematics is a human enterprise invented to describe the regularities seen in nature,[17] ignoring perhaps that their view that nature has regularities is itself a belief. There are scientists who are convinced that there is no such thing as mind, only the brain and its products. Other scientists believe otherwise: mind is a perfectly plausible construct that, for example, has utility when studying how people reason. There are scientists who are convinced that the brain is constructed of units that perform specific functions—this view informs much of fMRI research and its interpretation. Others believe that the many parts of the brain are networked and that any activity involves multiple actions and interactions among diverse areas. There is also considerable competition among scientists regarding scientific discovery. Examples include Joseph Priestley and Antoine Lavoisier in their race to discover oxygen,[18] the behavior of key players in the discovery of DNA,[19] disagreements among researchers currently investigating altruism,[20] and disputes over access to field sites among anthropologists.[21] Instances of data falsification deserve to be added to these examples.[22]

AUTHORITY

Further, scientific research, evidence, and explanations often acquire the status of authority. Certain findings and authors are cited far more often than others. From one perspective, this might be called wise science. In a way it is. What scientist wishes to cite other than the most authoritative findings or scientists? But it is also a manifestation of a hierarchical intellectual credential system.[23] Scientific publications that positively cite the work of other scientists tacitly acknowledge their authority for the moment and have the effect—perhaps unintended—of reducing *divides* among readers. Gould's maxim is a case in point. It is cited frequently as the basis for separating the domains of religion and science. His authority is recognized and well deserved for his many contributions in multiple areas of biology. Perhaps this is why his view of nonoverlapping magisteria retains its influence.

A related point is that both scientists and nonscientist audiences are sensitive to authority when cultural factors presage interpretation or, to recall the words of Francis Bacon, when we see what we believe. Studies designed to assess factors that lead people to believe what scientists say are in agreement with this point. Belief in what is said is not determined by whether it is consistent with evidence or its interpretation as endorsed by a national academy. Rather, people are more likely to see a scientist with elite credentials as an expert when he or she takes a position that matches their cultural values or estimates of risks.[24] A version of this process is currently in play among both scientists and their audiences over climate change.

The author of the study cited in the previous paragraph, Dan Kahan, recommends a strategy to offset the effect of cultural factors: "To make people form unbiased perceptions of what scientists are discovering, it is necessary to use communication strategies that reduce the likelihood that citizens of diverse values will find scientific findings threatening to their cultural commitments."[25] Taken at face value, this is a perfectly reasonable recommendation. However, such efforts have a long and unsuccessful history, particularly when applied to beliefs that many scientists have rejected, such as the existence of UFOs, ghosts, and much of pseudoscience. Or take the high frequency of unproductive discussions among scientists and creationists dealing with evolutionary theory. The communication-strategy recommendation overlooks an obvious point: it is belief, not communication strategy, that is at issue.

EMPIRICAL STUDIES

The matter of nonoverlapping magisteria might rest here except for the fact that domain-related overlaps develop when scientists and theologians study features of religion. What follows is a sampling of direct-evidence findings from empirical studies dealing with religion conducted using standard scientific method. Giving up one's religion is associated with a decline in health.[26] Religion-based transcendental meditation reduces the symptoms and signs of depression.[27] Pleasant afterlife beliefs are associated with better mental health.[28] Rituals, particularly specific types of praying, change the structure of the brain.[29] Religion aids well-being via social networks.[30] There is also the possibility that "the God to whom we pray" exists in the human brain as an ontological material category because it releases submissive behavior seen in the nonvocal aspect of petitioning prayer.[31] Participation in the package of behaviors that is a signature of many religions—belief, ritual, and positive socialization—reduces the aversive chemical and emotional effects of stress and uncertainty.[32] Compliance with religious rituals leads to an improved sense of social status and self-respect.[33] An increased sense of social status positively correlates with the density of dopamine receptors in the striatum, a region of the brain that plays a central role in reward and motivation.[34]

In effect, believing in ways that affirm religious beliefs leads to a decline in anxiety, normalizes aversive levels of brain chemicals, minimizes uncertainty, provides one with a specific place and status in the community, and may selectively release submissive behavior, which is associated with multiple positive health and psychological outcomes.

It is, of course, possible to argue that many of the preceding findings have nothing to do with God or an afterlife but rather reflect an "unholy alliance between medicine and religion."[35] Maybe all that is required to develop similar findings is a belief—perhaps that humanlike beings inhabit Venus—and to engage in nonreligious-based rituals, such as jogging, visiting the spa regularly, and eating healthy food. However, with the evidence currently in hand, a more plausible explanation is that religion-related belief and behavior influence believers' physiology, emotions, cognition, and behaviors in specific ways, and they do so more effectively than attending one's favorite spa, jogging thirty miles a week, and eating broccoli for dinner on Tuesdays and Thursdays.

OVERLAP—YES OR NO?

Further magisteria overlap is often present in the writings of scientists and scholars with a scientific bent. For example, it has been proposed that religious beliefs have their origin in the evolution of causal beliefs, which in turn had their origin in tool use.[36] Another proposal is that belief in God is deeply embedded in the human brain, which is programmed for religious experience.[37] That Christianity follows the laws of physics is yet another view.[38] Others argue that belief in religion has evolutionary adaptive consequences and that game-theory algorithms can explain religious responses.[39] Religious belief has been interpreted as reflecting a continuing ancient religious quest to which humans are committed, like it or not. That is, we have evolved such that we engage in this quest for which there is no alternative.[40]

Then there is the conjecture that religion is in our DNA. Francis Collins, the director of the National Institutes of Health for the United States, views it this way: "To see our species as embedded in a web of life and descended through natural selection from common ancestors with whom all life is shared is to see not human DNA but DNA *per se* as the prerequisite for the emergence of a life form capable of asking the question, What ought I do?"[41] For Collins, "What ought I do?" is a universal moral question and what religion is really about. Because the question is a product of our DNA, DNA can be viewed as the source of this fundamental moral law and religion.

Interesting speculations all. Each with wide or, at best, indeterminate *divides*. But do they differ from many speculations offered in the name of theology?

A MOMENT OF REVIEW

Let me review to be certain that key points are clear. Scientific speculations are eventually constrained by existing evidence and its interpretation. In contrast, much religion-based speculation is not faced with the same constraints. Asserting that there is a god or a heaven can and does occur without evidence that meets scientific criteria. Speculations about gods and heaven are only one facet of religion, however. For other facets, there is evidence consistent with the view that affiliation with a religion has definable and measurable physiological, anatomical, psychological, and behavioral effects. Both science and religion employ direct and indirect evidence as well as authority in their interpretation of evidence—for religion, both miracle cures and revelations are examples of direct evidence. At times, the two domains work together under the umbrella of scientific inquiry while addressing theology-related issues. Both attempt to narrow *divides*. Further, it is highly likely that there was a person named Jesus, that Matthew, Mark, Luke, and John were his disciples, that there was a Mohammad, and that some of the events described in the Old Testament are consistent with archeological and historical findings. By contrast, the evidence supporting much scientific conjecture is often flimsier. The likely presence of humanlike life elsewhere in the universe is an example. These speculations differ from religion-based views only perhaps because of the conviction among scientists that someday scientific evidence will be forthcoming.

CONCLUSIONS?

What conclusions might be drawn from this chapter? Many, perhaps, but the main one is this: the assertion that science and theology are nonoverlapping magisteria has more plausibility as a possibility about what might exist rather than a description of what does exist in the daily lives of scientists, believers, and theologians.

Still, there are important differences. At heart, scientific methodology is aimed at narrowing recognized *divides* through experimentation and precise statements about evidence. It acknowledges the presence of *divides*.

Often its views and findings can be tested. Often they change. That this happens daily is attested by the fact that yesterday's authoritative findings and authorities are not necessarily those of today. Theology too may aim at narrowing *divides*, but it is unclear if it has a methodology other than tradition and efforts to refine its interpretation of authority.

9
PHILOSOPHICAL CONSIDERATIONS

Where was Greg when I needed him? No doubt, I thought, wandering around the Piazza del Popolo or at one of the small restaurants on the Tiber side of the Spanish Steps eating ravioli, two of his favorite activities.

My e-mail went off with the hope of enlisting his help in dealing with the philosophy of belief. His response was not encouraging:

Dear Michael,

Sorry I can't be of much help. Simply too busy. Teaching duties here take up 60–70 hours a week, giving lectures, meetings with students, grading papers, and of course late dinners here and there with other faculty. And I've met a young lady, Francesca. Anyway, the real reason is that the philosophy of belief it's not an area in which I'm at home or know much about. You might contact Professors P or Q and see if they can help.

Professor P was cordial, brief, but not encouraging. "It is an extremely broad field not easily characterized. Philosophers have been arguing about belief for centuries and there's no consensus. This means that key questions still need to be asked or the whole enterprise has been poorly framed. You know of course that some philosophers doubt that beliefs exist—if they don't exist, they can't be studied. A thorough study would require several years and require several books. It's not something I can undertake."

I thanked him for his time and thoughts and departed.

Professor Q was less cordial and, at times, highly skeptical. I told him of my view that the brain was critical to study if belief and *divides* are to be understood. He disagreed. "The brain is unlikely to provide critical answers.

They [beliefs] are products of the mind and it's there where they reside and thrive. New information often changes them. At times we invent them. The brain may be essential for these events to happen but I strongly doubt that it will explain the content of beliefs, their meaning, or how new information changes them."

I also thanked him for his time and departed.

I spared Greg news about the outcome of his recommendations.

Floundering still, a past conversation with Mrs. X came to mind. Fortunately, I had recorded it in my notes.

"Where do you think your belief came from?" I had asked.

"I'm not sure, but I suppose I heard or read about it somewhere. Some real-life story by someone who eventually found her real parents."

"And how do you explain that it doesn't affect your relationship with your parents?"

"I don't know. At times I wonder if my brain isn't split in two, one part does the thinking; one part decides how to behave."

"The two parts aren't connected in your view?"

"Yes, that's the way it seems. I can imagine almost anything and it doesn't influence my behavior. After dinner the other night, I found myself thinking I was a Hollywood star and that people were seeking my autograph. Then I woke up and continued cleaning the kitchen, which I had been doing all along."

"Go on."

"At other times, I find myself behaving in ways I hadn't planned."

"For example?'

"Let's see. The other day I took my car to be fixed. I noticed that the mechanic had a political sticker on the shop wall. It wasn't a candidate I like. Soon I was involved in a heated argument with him over the merits of our candidates. I never do things like that, never once that I can remember."

Mrs. X had solved my problem. I would start with dualism and monism. "Philosophy, here I come," I thought.

A LONG HISTORY

For as long as there have been philosophers, there have been questions about beliefs. What are they? How are they created? Do they qualify as knowledge? Are they useful? How can they be justified?

Ancient Greece is a convenient place to pick up on their interest. A familiar group—Socrates, Plato, Aristotle, Heraclitus, and Empedocles, to name only a few—created an agenda of issues not only for their time but also for much of philosophy through today. During the ensuing millennia, Descartes, Hume, Locke, Berkeley, Spinoza, Bergson, Kant, Hegel, Russell, Heidegger, Lovejoy, James, Pepper, Wittgenstein, Quine, Fodor, and Dennett—again, to name only a few—as well as members of the church such as Augustine, Aquinas, and Cardinal Newman have grappled with the same issues, enlarged the agenda, and added amendments of their own.[1] Over the same period, members of so-called primitive societies have been engaged in similar activities.[2] Philosophy largely dominated inquiry until circa 1900. Since then, psychology and neuroscience have been active participants.

ABOUT THIS CHAPTER

Given the time involved and the brain power of philosophers, a formidable body of knowledge and theory has evolved. Thus, it is fitting to inquire about its relevance to belief and *divides*. I make no attempt to address the entire corpus of this knowledge, its complexity, or its many nuances. The modest aims of this chapter are as follows: to evaluate two philosophical views as they relate to belief and *divides*, to assess if and how they are informative, and to review selected findings from neuroscience.

First, it is worth discussing monism and dualism, two very different philosophical views with clear implications for belief and *divides*. I am interested in inquiring briefly as to why dualism persists. Monism, the view on which this book is based, is the next section in which the emphasis is on how monistic-based beliefs interweave with memory, situational belief, emotion, findings from neuroscience, and energy conservation by the brain.

As the chapter unfolds, recall that *divide* references how individuals—

not others—perceive the distance between a belief and the evidence they take as justifying the belief to be true. *Divides* may be narrow, wide, somewhere in between, or indeterminate. A narrow *divide* is present when a person perceives that a belief is strongly supported by evidence. A wide *divide* is present when a person perceives that there is no evidence supporting a belief. When a *divide* is indeterminate, a perceiver is uncertain if there is evidence that might justify a belief.

MONISM AND DUALISM

Much of what philosophers have discussed regarding belief and *divides* falls under the umbrellas of monism and dualism.

While the German philosopher Christian Freiherr von Wolff (1679–1754) first introduced the term *monism* in the eighteenth century, the view has roots extending back to well before the ancient Greeks. It is found in an early form in the writings of the pre-Socratic philosopher Parmenides. He argued that mind—the complex of elements in an individual that feel, perceive, think, will, and reason—and the physical brain are not separate and can be explained by one unifying principle or as manifestations of a single substance.

Identifying and characterizing the unifying principle or single substance are two of the tasks philosophers have embraced, and myriad attempts have been devoted to just these tasks. The results have fallen short of agreement, however. Monistic theories differ significantly in what is viewed as the unifying principle. It may be material, as in the doctrine that energy is the only reality. It may be spiritual, as it is for Hegel, in which mind or spirit is the true reality. For Spinoza, it is a substance or a deity of which the mind and body are attributes. Or it may be some mixture of mass, structure, energy, time, space, and information, a possibility to which we will come.

Dualism assumes that the mind and the physical brain represent two distinct and irreducible separate principles. Like monism, dualism takes many forms. For Plato, there is an ultimate dualism in being and becoming. For Kant, there is an ontological dualism between the nominal (that which is apprehended by thought) and the phenomenal or experienced world.

It also has a theological application as among the Manichaeans—a religion roughly contemporary in time with the Roman Empire—whose founder viewed evil in the world as resulting from an ultimate evil principle, coeternal with good.[3] For many neuroscientists, psychologists, and poets, there are variations of these views.[4]

DUALISM

Dualism and monism carry very different implications for these pages. The view taken here is monistic: essentially, what we usually refer to as "the mind" and "the brain" represent two ways of characterizing a single biological system that can be studied scientifically.[5] This means that a scientific explanation of beliefs and *divides* will be incomplete unless it is grounded in biology. For dualism, explanations of beliefs and *divides* pay minimal heed to their biological origins. Given the differences in these views, it might then seem that a further discussion of dualism could be dispensed with. Read on.

Although dualism has a history dating back to well before ancient Greece, it is most often associated with the French philosopher Rene Descartes (1596–1650) and what is arguably the most famous axiom in philosophy: "I think, therefore I am." For Descartes, the physical world is mechanistic and divorced from the thinking mind—that is, what we experience in awareness. There is perhaps no clearer description of dualism. In its strongest form, it allows for literally no interaction between the brain and its mechanisms and mind.

Descartes's assertion has had an influential yet controversial place in philosophy. It has been subject to serious criticism not only by philosophers but also by scholars from other disciplines. For example, the social psychologist George Herbert Mead views it this way:

> The unsatisfactory result of this division of nature between mind and the physical universe led to the objective idealistic systems in which nature was taken entirely into mind, not as a representation of an actual or possible reality outside of mind, but as the sum total of reality. . . . The undertaking failed, for one reason, because it identified the process of reality

with cognition, while experience shows that the reality which cognition seeks lies outside of cognition, was there before cognition arose, and exists independent of cognition after knowledge has been attained.[6]

Or, in the words of the neuroscientist Antonio Damasio:

What then, was Descartes' error? . . . Taken literally, the statement ("I think therefore I am") . . . suggests that thinking, and awareness of thinking, are the real substrates of being. And since we know that Descartes imagined thinking as an activity quite separate from the body, it does celebrate the separation of mind, the "thinking thing," (res cogitans), from the nonthinking body, that which has extension and mechanical parts (res extensa).[7]

In 1949, philosopher Gilbert Ryle characterized dualism as a "category error" and "the ghost in the machine." For Ryle, the death of dualism as a viable philosophy was only a matter of time.[8]

And, nearly two decades ago, psychologist Henry Plotkin summed up much of the criticism as follows:

What philosophy has done is refine the problems of knowledge without giving us answers that square either with ordinary life or with the extraordinary success of science itself.[9]

There is far more to Descartes's views than those that are considered here.[10] However to discuss them would divert attention from a key point: namely, that Plotkin's evaluation is spot-on. "Ordinary life" is the life most of us lead. In that life, the wisdom and concerns of philosophers often seem remote from what we experience and how we believe and behave—recall the earlier discussion about postmodernism. Nonetheless, dualism thrives. That it does clarifies in part why two-thousand-plus years of discussions about belief and *divides* have failed to provide compelling explanations about their biological nature and workings.

WHY DUALISM PERSISTS

We often experience things as if they are separate from our bodies. Such experiences are a type of direct evidence, and they are convincing and believable to those who experience them.

For example, close your eyes and imagine something as far-fetched as encountering a lovesick armadillo. You are likely to have a sense of an animal with an unhappy face. Now, imagine that the armadillo finds a mate. What you experience in awareness is likely to change to that of an animal with a happy face. Imaginings such as these flash into awareness and disappear rapidly—they can have a very short half-life. Nonetheless, with surprising ease, they can alter *divide* distance or confirm already-present views about how the world works. Unrealistic romantic afflictions, exaggerated self-appraisals, and the misinterpretation of events are often their aftermaths. Such imaginings don't necessarily require evidence. Nor is there the recognition that unperceived brain systems are involved in their creation. We are, as a number of scholars have put it, "born dualists."[11]

It's as if such imaginings are playthings—toys of the mind, so to speak—that can be created, manipulated, remembered or forgotten at will. And as we play, which we all do, there is literally no awareness that they might be contingent on the activities of unperceived brain systems. What we experience is disconnected from the body. Only at moments when a belief is associated with an emotion such as happiness, excitement, anxiety, fear, despair, disgust, or pain or when we are focusing on some part of our body is there an appreciation of the body's presence.

Further, we experience most actions as initiated in awareness. For example, you believe your car needs a new battery. Your mind seemingly initiates a plan to obtain one, and off to the store you go. Or you believe that it is necessary to review your finances, so one afternoon, you sit down with your spouse and discuss the family budget. Such experiences occur repeatedly as people attempt to make sense out of and manage their lives. Again, what is sensed is experienced in awareness as separate from the body; the detailed workings of the brain are hidden from awareness. There is nothing unusual about this hiddenness. The kidneys, spleen, liver, and immunological system operate twenty-four hours a day largely without any perception that they are at work.

Our perception that awareness is a system that is separate from our bodies thrives because it is consistent with our experience.

Our experience aside, however, it seems clear that our sense that our minds are separate from our bodies is untenable. Noting what happens when people are anesthetized, suffer from a severe concussion, or incur a serious brain injury should be sufficient to make this point: when the brain is not functioning, "I think, therefore I am" is nowhere to be found.[12] Nonetheless, and despite Ryle's prediction of its disappearance, a sure bet is that dualism will persist in daily life. That it will is a measure of the powerful influence of our experiences in awareness and the explanations with which they are associated. We may be able to imagine—even believe— that the brain is responsible for our perception that our minds are separate from our bodies, but that won't change what we experience. Such influence hints at one of the brain's seeming priorities: establishing beliefs is far more important than assessing how beliefs are created or justifying them.

MONISM

We return to monism. Five topics are discussed as they relate to belief and *divides*: memories of beliefs, situational belief, emotion, neuroscience findings, and energy.

MEMORIES OF BELIEFS

How are beliefs stored in the brain? One way—often referred to as *representation*—posits that a sentence or a symbol or a network of molecules is stored in the brain and may be accessed or recalled.[13] What we experience is consistent with this view as in instances when we are aware that we are actively searching our memory to recall a belief from an earlier time. In a manner of speaking, such memories seem to be hidden in a remote closet or in an attic of the brain waiting to be found following an active search. What we experience is something like this: we sense that, somewhere in our brain, there is a belief; we search to find it; suddenly it is recalled. In effect, one brain activity is to store memories, another is to search for them, and a third is to experience them when they are found. But unlike a physically tangible item stored in the attic, which, when found, is essentially the same as it was when it was stored, beliefs often undergo change while they reside in memory and are out of awareness. How this might happen, as well as its importance for understanding belief is a topic of coming chapters.

Situational belief. Situational belief is a version of the *extended mind hypothesis*, which deals with how external information initiates brain activity and changes in awareness.[14] For example, I am reading a book, and the telephone rings. A likely response is "someone is calling me": external information has triggered a change in my brain activity and awareness. Such experiences are similar to school reunions, where the faces and voices of once close friends initiate long forgotten memories of past moments.

Situational belief is so commonplace, so automatic, and so effortless that its occurrence usually is unrecognized. A sink full of dirty dishes, the smell of burning toast, an unexpected explosion, everyday conversations, and a cry for help are examples. Each is associated with a change of brain activity and awareness. Add to these examples the similar effects of the written word and a clear implication is that, except perhaps for sleep, the brain is continually influenced by information from outside the body. Such influence appears to begin before birth.

Research findings are consistent with these points. For example, the amygdala is a brain center that contributes to emotional states in recipients of external information. It is particularly responsive to the direction of the gazes of others when they send messages of anger or fear. Messages sent head-on are associated with significantly less amygdala activity and emotional response compared to the same messages sent when the sender's head is at an angle.[15] The degree of exposed eye white—squinting versus open eyes—also results in different degrees of amygdala activity.[16] Or, if one wishes to look to findings from studies of nonhuman primates, in certain species, submissive displays by subordinate animals directed toward dominant animals alter the brain chemistry of dominant animals.[17]

Situational belief explains the seeming paradox in which the absence of external information affects memory and initiates change in *divides*. For example, say you grew up on a farm, left the farm to attend college when you were seventeen years old, and haven't returned since. During your absence, the farm has been bulldozed away and now is the site of a housing project. Then one day you return to the site of the farm. Many of its details—exactly where the barn was located, whether its doors opened inward or outward, and so forth—can't be recalled. *Your ability to remember is contingent on the availability of specific environmental information that no longer exists.* Said differently, part of your personal history has been erased. You still believe that there was a barn and it had doors, despite the fact that memories of critical details are unavailable. You are unlikely to change your belief even though the *divide* about the barn and its doors has changed. It was narrow before you returned to the farm. Now it's wide.

Several points follow. One deals with how to characterize the brain. Anatomically, it includes the tissue within the skull and nerve extensions throughout the body. Operationally, it may be necessary to add information from the environment, at least during periods of wakefulness. Conceived of as an operating system, the brain includes brain tissue, nerves throughout the body, and information from a variety of sources, including the environment. There appears to be little difference between a response associated with the pain of a stubbed toe or the irritation that accompanies the recognition that one's car has a flat tire. It follows that the exclusion of environmental information from efforts to explain how the brain works invites serious review.

A second point deals with belief. External information can trigger beliefs. For example, visualizing a sudden movement in high grass may be followed nearly instantly by the belief that a snake is present.

A third point deals with *divides*. *Divide* distance varies with external-information type. For example, a difficult-to-interpret noise or smell often leads to the belief that something is amiss with a sense of uncertainty about its cause. There is an absence of evidence that can be quickly and easily interpreted. *Divides* are likely to be indeterminate at such moments. On the other hand, if you hear a familiar voice, such as that of your child, you will believe that your child is its source and there is no *divide*. Because environmental information is endless in its variety, there is literally no limit to the number of its *divide*-related influences on brain activity and belief.

EMOTION

There are few more-daunting areas of scientific research than that of human emotion. Part of the reason is that it is difficult to study. Awareness of an emotion may range from the barely perceivable to an extreme that dominates one's awareness and behavior—recall my experience of reading about jaguars and water (see chapter 6). At times, two or more emotions combine, as in moments of fear and disgust.

Academic research has focused on cognition far more than emotion and, until recently, they have been viewed largely as separate brain systems. Studies over the past two decades have altered this view significantly. They have shown that the brain areas responsible for cognition and emotion are connected via integrated neuronal networks and that they often "compete for dominance" in the interpretation of information.[18] For example, the amygdala is linked to the prefrontal cortex (which is associated primarily with cognitive activities such as social choice, predicting future events, planning behavior, and controlling emotion), the orbitofrontal cortex (which is associated with sensory integration and decision making), and the anterior cingulate cortex (which is involved in attention, motivation, and error detection).[19]

The study of emotion has been complicated further by the presence of competing and often-nonoverlapping explanations of its origins and effects. For example, the James-Lange theory posits that emotions are largely a consequence of bodily change.[20] The evolutionary approach stresses that emotion has evolved to serve particular adaptive challenges.[21] The neurobiological approach attempts to integrate emotion and features of cognition.[22]

Clearly there is much more to learn. Yet already a great deal can be said about the influence of emotion on belief and *divides*. For example, consider the following paragraphs:

Emotion has a powerful effect on belief, particularly when there is a high level of emotional arousal, which leads to attention narrowing and an enhanced focus on experience.[23] That is, the brain may develop beliefs in which emotion is the principle organizer of information. Beliefs associated with disasters such as 9/11, frightening experiences such as being threatened by an animal, or unexpected pleasurable outcomes such as a surprise party in one's honor illustrate this effect.

Emotion influences decision making via visceral reactions, such as disgust, and during the anticipation of fear, anxiety, and uncertainty.[24] People put off visiting the dentist and undertaking painful jobs because of the displeasure they anticipate, just as they avoid repeating behaviors that disgust them.

Beliefs associated with pleasure and reward are favored over those associated with negative emotions often irrespective of evidence: believing that one's lost child will be found is a more tolerable and pleasurable belief than its alternative.[25]

Culture, belief, and emotion interact. Some emotions, such as anger, fear, and sadness, appear to be universal and independent of cultural differences.[26] Others are influenced by culture and context, such as responses to events among members of collectivist and individualistic cultures.[27]

The upshot of these examples and a host of related findings is that belief and *divides* may be influenced as much if not more by emotion than by cognition.[28]

Do emotions widen or narrow *divides*? They appear to narrow them when there is a high level of positive emotional arousal, which glues together emotion, experience, and belief—in effect, the physicality of emotion appears to counter doubt. For decision making, both positive and negative emotions close *divides*—there is literally no *divide* when one decides not to order one's least-favorite food for dinner. Culturally influenced emotions may either widen or narrow *divides*. When an emotion is concordant with shared cultural views, narrowing is likely. Widening is the rule in the absence of concordance.

NEUROPHYSIOLOGICAL STUDIES

It may seem surprising that at this moment in time, it is necessary to assert again that without a brain, there is neither belief nor *divides*. Yet this is the case. Thus it is not unexpected that neuroscience studies have been and continue to be rich sources of findings and insights about how the brain works.

For example, in studies in which research subjects are asked to think about something they believe or disbelieve, fMRI technology reveals differences in the areas of the brain that are activated. Belief activates the region known as the ventromedial prefrontal cortex, which has an anatomical link with areas responsible for the cognitive aspects of belief, emotion, and reward. The areas activated during disbelief include the limbic system and the anterior insula, brain regions associated with awareness of visceral sensations, such as pain and disgust, and in negative appraisals of sensations dealing with taste and smell.[29] Other studies point to the involvement of the temporal lobe in belief-related reasoning.[30] Still other studies suggest that the frontal lobes are active and that parietal lobes "power down" (are less activated) during intense prayer, which is a form of belief.[31] To believe in a spiritual sense is associated with decreased activity of the right parietal lobe.[32] And moral judgment can be altered by disrupting specific brain regions.[33] What these examples share in common is that specific areas of the brain are activated during states of belief and disbelief.

There is also physiologic evidence. The effect of excessive alcohol consumption is perhaps the most familiar example. Elevated brain levels of alcohol lead to physiological change in the brain and body and impairment in detecting performance errors—that is, how one is performing on a test.[34] At times, alcohol consumption is associated with the appearance of novel beliefs that later are lost to recall—"Did I really say that?" Fatigue has similar effects. There are drugs such as psilocybin that initiate mystical-type experiences that are completely novel to the user and may have substantial and sustained personal meaning and spiritual significance.[35] Antidepressant drugs sometimes change people's beliefs about their personal worth. And elevated levels of the hormone oxytocin intensify the belief that others are trustworthy.[36]

The preceding examples comprise only a small sampling of findings and insights from the neurosciences. Nonetheless, this much seems clear: the activation of multiple areas of the brain during different facets of belief, when combined with the effects of physiological alteration and environmental information on brain systems, suggests the following are highly likely. Beliefs and *divides* are products of multiple brain systems. These systems perform specific tasks. Neural networks connect and communicate between systems. Systems are responsive to chemicals that facilitate or constrain their operation. Energy costs to the brain affect the workings of systems.[37]

ENERGY CONSIDERATIONS

One place to start with a discussion of the energy cost of brain work is by asking why we experience "thinking" about information that we can't explain easily as so strenuous. Familiar experience suggests an answer. It is not easy to organize a large wedding, design a novel software program, devise an intricate strategy in a war effort, or evaluate the relevance of another's belief when there is incomplete or conflicting evidence that is susceptible to multiple interpretations. Not only is brain energy expended in organizing and making sense of information, but it is also involved in maintaining one's cognitive focus, dampening the intrusions of external information, and controlling the influence of emotion. These are energy expenditures in addition to the default energy requirements of the brain. Thinking is hard work because it is an add-on cost to these ever-present default costs, which are estimated to range between 60 and 80 percent of the brain's energy budget.[38] This perhaps explains why occupations that require extended periods of thinking, such as managing a multifaceted company, are so stressful. The degree of stress is a proxy for the high amount of energy required by the brain to think. Perhaps, too, energy requirements explain why people often respond with irritation when their deeply held beliefs are challenged: they have to expend energy responding to the challenge. Dismantling or changing beliefs can be costly to the brain and emotionally unpleasant.

The energy costs associated with serious thinking can be contrasted with those of the *wandering brain*: without obvious effort, direction, or stress, beliefs, memories, and emotions appear and disappear from awareness.[39] This often happens while driving a car along a familiar route or when one is engaged in a familiar activity. The brain seemingly wanders aimlessly, as suggested by the presence of unconnected memories, beliefs, and emotions that flit through awareness. Such moments occur without recognition of much of the passing landscape or a sense of time, although obviously there is a part of the brain that is managing the driving.

While we experience the energy costs of thinking as high, often they can be reduced dramatically by belief. An emotional twinge or an environmental cue may be sufficient to initiate a belief and thereby bypass the energy expenditure of thinking. Beliefs can be quick and automatic, as during moments of intuition. Minimal thought may be involved. They can often override ambiguity and uncertainty as well as preserve energy.

For example, today's news might report that Chinese spies have been arrested in Tokyo, there's a decline in Japan's stock market, Japan is objecting to the continuing presence of United States armed forces on its mainland, a major Japanese automobile company has declared bankruptcy, the yen has declined in value, and there is evidence of government-based corruption. Efficient as the brain often is in processing information, it can't easily organize and make sense of much of what it receives, especially if evidence is indirect. A quick and efficient way to manage such information, establish a narrow *divide*, and reduce energy expenditure is to package such information into a belief, such as "Japan is in a state of turmoil."

Why packaging occurs is of interest, of course. One possibility is that there are adaptive advantages associated with decreasing the energy costs of trying to make sense out of ambiguous and uncertain information. Ambiguity and uncertainty are known to be stressful and lead to chemical changes in the brain as well as aversive body states. Reducing their effects may preserve energy for more critical brain tasks.

SO HOW ARE PHILOSOPHICAL CONTRIBUTIONS TO BE JUDGED?

There is no requirement that philosophers explain much of what people experience in daily life. As with all areas of knowledge, philosophy has its own agendas, unresolved perplexities, and favorite topics. Still a rough assessment regarding its contribution to the understanding of belief and *divides* is possible. To try and explain beliefs and *divides* in ways that are consistent with the mind-body assumption of dualism would seem to limit inquiry primarily to what we experience. Historically, this approach has been unproductive largely because the operations—systems—of the brain are hidden, often seemingly far from rational, and usually excluded from explanation. Characterizations of how the brain (or mind) works are primarily descriptive and don't address the mechanisms responsible for its operations. On the other hand, monism is open to trying to explain how a single system can account for beliefs, *divides*, awareness, memory, and so forth. To do this, monism will have to address a variety of issues that it has largely avoided, the most important of which deal with awareness and how the brain processes information.

10

AWARENESS, BELIEF, AND THE PHYSICAL BRAIN

By training, I'm a psychiatrist who has worked most of his life in the field of neuroscience. Thus it might be expected that it is there that I would have begun my inquiries. Because of my familiarity with the field, I resisted as long as I could. Why? I wanted to find out what others had discovered; their work deserved a hearing. They had addressed many of the same issues with which I was struggling.

Psychologists provide insights about how information is manipulated and managed. Historians are informative about the uses and durability of beliefs and, often, how they are slow to change even when there is strong contradictory evidence. My venture into evidence and its sources, while revealing, leaves many questions unanswered. Computer experiments confirm, once again, that we impose our beliefs on evidence and configure it in ways that are consistent with our beliefs. A look at religion and science suggests that there are significant overlaps in what is viewed as evidence and how it is interpreted. The findings of philosophers can be interpreted from several perspectives, but one thing is certain: philosophers hold very different views about the relevance of the brain in explaining belief.

I phoned Greg and told him that a look at neuroscience was next on my agenda.

It was clear he was uneasy, "You know I'm skeptical about science."

"Yes, but go on," I insisted.

"It's for some of the very reasons we've discussed. Science is always evolving—its findings change from day to day. The brain has hundreds if not thousands of systems for carrying out its tasks, and billions of neurons are involved in these tasks. And . . . and only a very few of these systems are well understood. There are simply too many unknowns. Isn't that what you've been preaching?" He paused.

"In a way, but still, go on," I suggested.

"It's like this. You'll be attempting to answer questions where you are uncertain about the validity of much the evidence, and the evidence is constantly changing. And, of course, often there are different interpretations of evidence even among your colleagues."

Clearly irritated, I interrupted: "Are you saying that nothing can be said unless everything is known? If so, that's a form of intellectual nihilism. Anyway, there's another view. Science is about solving puzzles. Granted, complex puzzles are not solved all at once, but it doesn't follow that even partial solutions should be ignored."

"Let me think about it," he replied. "I'll phone you on Tuesday morning at your home—OK?"

Tuesday morning, the telephone rang. It was Greg. "I've been reconsidering, and I think you're right. Even with partial information, some things can be said. Otherwise, as you say, nothing would be said. Just one caveat, please: warn your readers that much of what you are saying is speculation."

We finished the conversation with a discussion about his newly found girlfriend, Francesca, and my reminder that his change of heart was an example of his changing a belief. He laughed and signed off with "*Arrivederci.*"

Greg's skepticism is not without merit. There are many unknowns. This was and remains the case for the high activity of brain serotonin in dominant male vervet monkeys. And he is right about the complexity of the brain, which, given current knowledge, couldn't be addressed in complete detail short of perhaps twenty thousand pages (which would be out of date before the pages went to print). Thus, only selected brain systems and operations are addressed. For example, the capacity of different systems to process the same information in parallel and the effects of both development and ageing on these processes will be overlooked.

Despite Greg's concerns, I was optimistic. As I turned to the neurosciences, I anticipated that delving into the workings of the brain would lead to new insights about the how and why of belief and the sources of *divides*. Relevant evidence is available and it invites explanation. Answers to questions might be only moments away.

WHAT CAN'T BE DONE

It is an axiom of neuroscience that a detailed understanding of how the brain works cannot be achieved by analyzing states of awareness. A second axiom is that the understanding of the detailed workings of the brain is still very much in its infancy and changes daily in response to new research findings and theories. I agree with both of these axioms.

Still it is possible to combine our awareness of our beliefs and *divides* with available evidence documenting the brain's unperceived systems and consider how these systems and operations might inform what we experience in awareness. Given the axioms above, it follows that any such effort will be speculative. That is the case here.

AWARENESS

Two quotations from neurologist Robert Burton's 2008 book, *On Being Certain: Believing You Are Right Even When You Are Not*, provide a convenient point of departure:

> The message at the heart of this book is that the feelings of *knowing, correctness, conviction, and certainty* aren't deliberate conclusions and conscious choices. They are mental sensations that *happen* to us.[1] . . .
>
> To be effective powerful rewards, some of these sensations such as the *feeling of knowing* and the *feeling of conviction* must feel like conscious and deliberate conclusions. As a result, the brain has developed a constellation of mental sensations that feel like thoughts but aren't.[2] [Burton's italics]

In effect, what we perceived as beliefs and *divides* are "mental sensations" that not only seem to exist on their own during moments of awareness but also seem unrelated to the systems and operations of the brain, which are their source. Further, our perceptions that we are making decisions and that they and beliefs cause behavior are also illusory.

That said, we encounter an interpretative conundrum: if our perceptions are illusions, speculations based on perceptions will also be illusions.

However, if states of awareness are viewed simply as mental sensations and we discard our illusory interpretations, then perhaps our speculations about the brain's workings may be less illusion influenced.

A further point. At present, we have no way of reliably measuring beliefs or other contents of awareness. It is this inability that in part makes them so difficult to study. Nonetheless, we do know of beliefs and *divides* because we are aware of them and they can be described. Awareness then is an obvious place to begin taking a closer look at Burton's views.

Awareness is the quality of being aware of something within oneself or of an external object or event.[3] The definition squares with what we experience. We are often aware of ourselves as separate from others, parts of our bodies, imaginary things, thoughts, emotions, actions, beliefs, *divides*, moments of self-control, making decisions, and more—the list is long.[4] What brain systems and operations might account for these moments of awareness? Four examples follow.

1. *Belief as a cause of an action.* You are driving down a road and a dog unexpectedly darts in front of your car. You apply your foot to the brake, swerve the car to the left, and avoid hitting the dog. Attention and action seemingly flow without effort.[5] This is an example of an *automatic response*, a response in which unperceived brain systems initiate action prior to your awareness that you have acted.[6] The temporal sequence of events runs something like this: your brain perceives the dog, it initiates a response, the response is followed by your awareness—your mental sensations—that you perceived the dog and acted to avoid hitting it.[7]

Beliefs and *divides* are seldom experienced during automatic responses largely because of their short duration. However, they appear frequently following such responses. For example, you can remember what you experienced: you recall seeing the dog, pressing the brake, and swerving the car. You are likely to believe that you perceived the dog and initiated your response to avoid hitting it.[8] Because the event is over and you missed hitting the dog, the *divide* separating your belief from what you recall—your evidence—is likely to be narrow.

It might seem from this example that people can't know what they are doing. But that isn't entirely the case. You do know what you did almost immediately after acting. It's just that the belief that you were aware of your actions prior to acting is wrong. On this point, the evidence is clear: studies consistently show that neuronal activity *precedes* awareness of the intention to act or an act.[9]

Your response to the dog is not a random behavior. Automatic responses are dependent on the presence of transparent models (sometimes referred to as templates or plans) for dealing with familiar situations. In this example, if the model is mostly innate, a first-time driver would respond essentially the same way as an experienced driver. Recall that people don't have to undergo extensive learning to withdraw their hands from a hot flame. They do it right the first time. On the other hand, if the model is mostly learned, the first-time driver is likely to hit the dog, do nothing, panic, or wreck the car.

Automatic responses are exceedingly common. For example, you drop something fragile and you "automatically" act to catch it before you are aware of your behavior. Or you experience an itch and your hand starts to move to scratch the itch before you are aware of the movement. Or you trip while walking: as you fall, your response to prevent hurting yourself begins before you are aware that you are responding.

Models are critical to the scenario developed in this and the remaining chapters. They are discussed in detail in chapter 14. For the present, the important points are as follows: Unperceived representations of models appear to exist. They are stored in the brain. They are triggered by internal and external information.[10] They can initiate or bias action.

2. *Belief that a thoughtful decision is being made.* You go to a restaurant for dinner. Three entrees are offered: fish, meat, and pasta. You have eaten each before. As you review the menu, you sense that you are making a decision about which item to select. A decision is being made, but not in awareness.

In this situation, midbrain dopamine systems associated with pleasure and reward appear to be activated by environmental cues—menu items— each of which is associated with a different level of activation of pleasure and reward systems.[11] As you deliberate about what to eat, unperceived neural correlates of mental rehearsal precede decision and action—that is, the brain is planning ahead and evaluating actions before one is aware of pending actions or before they are initiated.[12] If, in the past, the taste of fish was more pleasurable than that of pasta or meat—that is, of the three menu items, fish is associated with the greatest degree of activation of pleasure and reward systems—you are likely to select fish for dinner. Your past reward and pleasure history predicts your behavior.[13]

Should your dinner partner ask why you selected fish, you are likely to reply that the last time you ate pasta and meat, you didn't like them, but you do like fish. As with avoiding hitting the dog, your belief about your choice occurs after the choice of fish has been initiated. Nonetheless, your answer is likely to be a reasonably accurate way of characterizing events in the brain that occurred out of awareness. This example invites a revision of the view that beliefs sometimes come first and explanations for beliefs come second.[14] At least three steps are involved, not just two: (1) a stimulus initiates a belief, (2) awareness of the belief occurs next, and (3) awareness of searching for explanations follows.

How you perceive *divides* in menu-type (multiple-choice) situations varies. If, in the past, your experiences with pasta and meat were unpleasant, there is sufficient direct evidence to narrow *divides* for either choice: you believe that they are unpleasant-tasting foods and you have evidence justifying your view. The *divide* will also be narrow for fish: you have memories of its pleasant taste. On the other hand, if you haven't tasted any of the items on the menu, the *divide* for each of the entrees will be indeterminate.

3. *Belief that you have planned a trip.* You have a holiday from work. You wish to take a vacation, so you go about planning it. Your choice is to visit a park in Colorado. A day goes by and you alter the plan: you will include a visit with a friend who lives near the park prior to visiting the park. Another day passes and you alter your plan again; this time, you will visit your friend after visiting the park. Following several more revisions, the details of the trip seem finalized. You telephone your friend and confirm that you will arrive on a specific date.

Through all of this, you believe that you are making decisions about the trip. But, as with the previous examples, your belief follows on events taking place outside awareness: as noted, the brain plans and evaluates actions before one is aware of the pending actions or before they commence.[15]

This explanation invites several questions. First, why is the park in Colorado selected? External information, such as reading a travel magazine, is one possible influence. It could have triggered your interest—this would be an example of situational belief, discussed earlier. A second possibility is that the brain is wandering, and memories, beliefs, imaginings, and emotions are experienced in awareness without your sense that you are initiating them.[16] For example, at one time in the past, you may have wished to visit the park but were unable to do so. You didn't recall this wish when you started planning the trip. But at a moment during planning, your brain wanders and your interest in visiting the park seemingly appears from nowhere.

Second, why is the park selected in preference to alternative sites? In a process similar to selecting an item from a menu, the activation of pleasure and reward systems is likely to be part of the answer. You might have been to the park before and enjoyed it, in which case, there are pleasurable memories. Or you might have been to a similar park. Pleasure and reward are unlikely to be the only influencing factors, however. Pragmatic models that deal with details of the trip may influence your decision. Such models are based on past experience and secondary evidence. In this example, a pragmatic model would address factors such as the time required to drive to the park, the energy required for activities such as hiking, and the expenses of the trip. Unperceived competition may occur between pragmatic models and pleasure and reward systems: the pleasure and reward systems may be activated, but if the pragmatic model predicts excessive costs and very little available time to spend in the park, the trip may be cancelled or modified.

Third, how does visiting your friend enter your plans? A possible answer is that the location of the park and the location of your friend who lives near to the park are connected in the brain. It's easy to demonstrate this possibility: close your eyes and think of a city and note what follows in awareness. Then pick another city and compare what follows in awareness with what appeared for the first city.

4. *Free will.* Is there free will? That is, is there voluntary choice or decision? Answers to this question are far from clear.[17]

Consider the following example. You have inherited a considerable amount of money. You wish to invest it, but you have minimal experience in financial matters. From a variety of sources, you obtain recommendations about investing. With each new recommendation, you experience a change in your evolving plan about how best to invest. Last week, you were thinking about buying stocks, but today, an annuity looks most interesting. After a while, there are multiple options. Throughout all of this, you believe you are involved in making a voluntary choice based on the recommendations you have received.

In this example, unperceived brain systems appear to be creating and revising largely learned models in response to new information. As new models are developing, they are influenced by memories of prior behavior, pleasure, reward, pragmatics, and a variety of other factors, such as one's age and sex, health, and social status. A likely critical factor in the revision process is the brain's rehearsal of possible outcomes for each version of a model. This process appears to combine unperceived neural rehearsal for possible future events and "preplay": "internal neuronal dynamics during resting or sleep organize parts of the brain into temporal sequences that contribute to the encoding of related experiences in the future."[18]

Eventually, the recommendations for how to invest your money cease. You invest in gold, something you hadn't considered when you began your search. A free-will choice has occurred because there have been extended periods of awareness associated with creating, revising, and assessing a variety of models and rejecting all but one.[19]

Free will can be put another way. Choices aren't random. They don't occur without the presence of models or some equivalent. In this example, the models deal with investment choices and possible financial outcomes. *What has happened is that you have selected a model from those available in your library of models.* But note that your choice is limited to models in your library. The choice is "free" in the sense that you might have opted for investing in bonds rather than gold.

Throughout this process, you are likely to believe that you are evaluating different investment possibilities. You are, but, again, your belief temporally trails unperceived evaluation and decision-making processes. You may also believe that you are reviewing models and rejecting some because they don't seem to promise sufficient financial return. This also is happening.

At least two factors appear essential for free will. One is the availability of models from which to select. The second is the presence of systems that can prioritize models to facilitate selection. Systems dealing with pleasure and reward and pragmatics have been mentioned as facilitators. But as will be discussed in coming chapters, there are other possibilities.

The four examples are consistent with Burton's characterization of awareness: "The feelings of *knowing, correctness, conviction, and certainty* aren't deliberate conclusions and conscious choices. These mental sensations *happen* to us."[20] What has been added here is an emphasis on the timing of events in the brain and a discussion about possible brain systems that might account for them.

If the beliefs and *divides* we experience in awareness are illusions, are there significant consequences? From the perspective of daily life, there would seem to be few—after all, you did miss hitting the dog, select fish for dinner, visit a park and your friend, and purchase gold. Ideologically, however, consequences can be anticipated. These will deal with how humankind conceives of itself and the importance it attributes to awareness as the initiator of choice.

THE PHYSICAL BRAIN

The brain's physical properties are those of material-energy-space-time reality. They promise to inform our understanding of beliefs, *divides*, and awareness.

The brain has systems that are sensitive to environmental information, such as noises, smells, light, and faces. It has connections throughout the body that transmit information about touch, temperature, pain, and location. It has systems that initiate action.[21] It has systems that perform information-processing tasks, such as spatial representation, causal modeling, belief creation, and belief rejection. It creates and stores representations of memories, beliefs, and models. Many of these systems are involved in the creation and management of beliefs and *divides*. Which systems are involved varies in part with the type of belief and *divide*.[22]

Events in the brain occur in a chemical-and-electrical milieu. There are cells that transmit electricity via releasing molecules that initiate electrical activity in connecting cells. Our current understanding of the brain suggests that representations of beliefs, *divides*, memories, and models reside in this milieu, which consists of cells, their connections, and their surrounding environment. The chemical nature of cells means that molecules are involved (recall that many of the brain's molecules are affected by other molecules, such as those of drugs). Molecules have weight and structure. That they do means that they are not infinitely flexible.

Viewed this way, two uses of *belief* and *divide* have been introduced. *One use refers to the beliefs and divides that we experience in awareness. The other use refers to their unperceived representations in the brain.* Although it is uncertain if what we experience in awareness conforms to or can be interpreted using the laws of physics, events in the chemical-and-electrical milieu follow these laws.[23]

The default state of the brain appears to be one of continual electrical-chemical activity so that there is sufficient energy and metabolism for the preservation and management of tissues responsible for systems, memories, beliefs, models, and the like.[24] For example, during resting states in both humans and nonhuman primates, there is metabolic and electrical activity throughout the brain, and specific areas exhibit highly organized patterns of activity.[25] Further, the resting brain appears to recapitulate activity patterns that occurred during recent experiences.[26] In effect, not only is the resting brain active chemically, metabolically, and electrically, but it also appears to be involved in internal housekeeping and information management, such as belief and *divide* consolidation and revision, as well as preparation for the future.

LIKELY ANSWERS

What can these systems suggest about beliefs and *divides* and awareness? The likely answer is that *what we experience in awareness as beliefs and divides reflect representations stored in the brain.*[27] Two possible ways this might happen will be considered with the caveat that there is no compelling evidence to support either way.

One way is analogous to changing the luminescence of a variable-intensity lightbulb. In its resting state, the energy level of representations of beliefs and *divides* is low and they are not experienced in awareness. In an energized state, beliefs and *divides* are experienced in awareness (cartoonists may envision this possibility when they draw a lighted lightbulb above the head of a person having a thought). *Our awareness of beliefs and divides thus may be due to changes in the energy state of representations in response to internal or external information.* This possibility might be termed the *activation hypothesis.* One of its attractive features is that in doesn't require a transformation of the information in representations to accompany energy-state changes.

A second possibility is that there is a transformation process in which beliefs and *divides* experienced in awareness reflect alterations of their representations. As mentioned, there is not a great deal of evidence in support of this possibility. Nonetheless it is plausible, perhaps more so than the activation hypothesis. (If I had to guess, transformation would be my choice.) It does, however, have to address the fact that molecules are not infinitely flexible and electricity has specific qualities that may determine the types of transformations that can occur as well as affect what is perceived in awareness.

Let me pause a moment and review. What is being suggested here is this: On one hand, we experience beliefs and *divides* in awareness. They are illusions in that they are not initially consciously responsible for what we experience or the basis for decisions or actions even though we perceive them this way. On the other hand, there are representations of beliefs, *divides*, and models that reside in the material brain. These representations continually undergo revisions and updating. They are related to what appears in awareness. The degree to which what is experienced in awareness accurately reflects these representations is unknown, although there is no reason to assume that there is a great deal of difference between representations and what is reflected in awareness.

Behavior is initiated or biased by unperceived representations, not the contents of awareness.

A WARNING

Readers should not depart from this chapter assuming that there is consensus regarding many of the points that have been discussed. For example, for many authors, states of awareness (or states of consciousness) are not passive states for which unperceived brain operations are responsible. Rather, they are viewed as having influence or control over selected brain operations and behavior.[28]

INFORMATION-PROCESSING OPERATIONS

Information has many definitions and ways in which it is conceived.[29] The one adopted here is as follows: information = energy that alters brain activity. It could be defined in terms of other measurable features as it is in information theory.[30] Only the energy definition is used here. It is essentially synonymous with the usual meaning of *stimulus*: something that rouses or incites to activity.

Viewed this way, there are both internal and external sources of information. For example, an unexpected but familiar sound is usually sufficient to initiate awareness of the noise and belief about its possible cause. Such responses are consistent with studies indicating that both external and internal information are forms of energy and alter the probability of the contents of awareness. There are also well-studied examples of *mirroring*, a state in which neuronal activity in an observer's brain closely corresponds to the neuronal activity responsible for the behavior of the person or animal being observed—in effect, an observation initiates a change in an observer's brain activity (see chapter 13).

Applying these points to beliefs and *divides*, likely effects of information include: initiating the processing of both external and internal information; initiating the creation and storage of beliefs, *divides*, and models; reorganizing representations of beliefs, *divides*, and models; activating pleasure and reward centers; initiating belief and *divide* representations that are reflected in awareness; and initiating behavior prior to awareness that the behavior is occurring.

Other possible brain features and systems are discussed in coming chapters. For this chapter, selected key points are these. The brain has multiple, unperceived information-processing systems. They influence beliefs and *divides* as we experience them. For example, preplay and event rehearsal narrow *divides* in advance of the awareness of a belief. When models are involved in free will, choices are limited to a person's available models. And it is information (energy) that initiates unperceived systems and operations of the brain that are reflected in awareness of beliefs and behavior.

Finally, we are recipients of a brain that does marvelous and strange things. Only a very small percentage of its many possible operations have been addressed in this chapter. It's unlikely that the brain has evolved the way it has on purpose. More probably it is the product of multiple influencing factors stretching back millions of years. This raises the possibility that an informative approach to exploring these marvelous and strange things is to consider their possible evolution. That is where we are headed.

Greg's e-mail was encouraging.

I like this chapter, it's very informative and need I mention that it will invite criticism. Details to follow.

Best,

Greg

11

THE BIOLOGY OF BELIEF

Our e-mails read this way:

Dear Greg,

Thanks for your comments on the Awareness chapter. I was also pleased with it. And yes, it will invite comments.

Turning to other matters, things are fine and as usual here. Have the next chapter in draft. Do you have time to comment on a half-page summary of my approach?

Hi Michael,

Still very busy with teaching and, as you might suspect, more time with Francesca. Of course, send your summary.

Dear Greg,

My approach runs something like this:

1. The brain systems discussed in the awareness chapter as well as an immense amount of data from history, psychology, behavioral observations, and medical science need to be given a framework.
2. Evolutionary theory seems the obvious choice. Why? In part because it's the framework most consistent with my focus on the brain rather than the "mind." In part because the brain systems and the brain's bias to believe didn't appear overnight. They are the likely consequences of multiple past

events and selection processes. Presumably many of the systems were selected because they provided an adaptive advantage, which they may still do, although this is arguable for some systems.

3. Details of past events possibly influencing selection along with their consequences would follow.

Your thoughts, or do you want me to guess?

Hi Michael,

One of the reasons I enjoy our friendship is because we think differently. And we certainly do when it comes to evolutionary theory. Perhaps you are comfortable with it, but I'm not. I'll grant that it's the best theory around to explain a host of points you're discussing. I'll also grant that there is no compelling alternative theory of which I'm aware. Still, I'm uneasy. It's over-weighted in speculation relative to the available evidence and many of its ideas can't be tested. From what I can see there are internal inconsistencies, which often lead to different ways of making sense out of the same evidence. To me, this suggests that the theory lacks precision. Put another way: yes, points can be explained but with what degree of certainty?

I'd like to be more enthusiastic. But let me be clear on one point: I can't argue that there is a better framework. If you proceed as your summary suggests, which I suspect will happen, I will be delighted to savor your effort.

Best,
Greg

As usual, Greg had a point.

WORDS OF CAUTION

Insights dealing with ways in which beliefs and *divides* might have evolved as preeminent features of the brain may emerge from attempts at reconstructing their histories. This and the following four chapters engage in this task.

But first, some words of caution.

Efforts to reconstruct the past, especially the very distant past, are perilous and beset with myriad potholes. Evidence that might nail down critical contexts, events, or behavior frequently is unavailable. Theoretical biases may overvalue specific areas of inquiry and ignore others. For example, among paleoanthropologists, there is a notable absence of inquiry about the reproductive strategies of our ancestors.[1] Then there are disagreements among scholars over the authenticity and interpretation of findings. There are also vexing and unavoidable asymmetries in evidence. For example, chemical methods used to date the past age of physical evidence (pots, dirt) and biological evidence (bones, hair, DNA) can be quite accurate. However, very little may be known about the behavior of those living at the times that coincide with the chemical dates.[2] Thus, *divides* often are indeterminate when trying to tie together chemical-based evidence and behavior.

Another factor is that research reports documenting the discovery of ancient humans and other hominids along with reevaluations of previously reported findings are published weekly.[3] In turn, today's reconstructions may be obsolete tomorrow. An example deals with the morphology of *Homo erectus*, one of our ancestors, which is currently undergoing review as a result of new evidence and improved analytic techniques.[4] Recently, the discovery of remains of modern-day humans living in China one hundred thousand years ago has been reported.[5] Few scholars suspected this possibility ten years ago. Fifteen years ago, few experts would have suggested that at least twenty-five different hominid species have evolved during the past six million years.[6] Nor, until recently, did anyone predict that approximately 1.2 million years ago, the ancestors of today's humans were possibly an endangered species with fewer than twenty-six thousand individuals capable of breeding.[7]

REVERSE ENGINEERING

Much of reconstruction involves reverse engineering. This amounts to extrapolating back in time using present-day findings and methods. Studies documenting the rate of genetic change from specific periods in the past to the present are examples.[8] Or the same brain chemicals may be found among today's humans and chimpanzees, which split from a common ancestor millions of years ago. A reverse-engineering inference might be that the chemicals were present in their common ancestor prior to the time of the split. The inference could be wrong, of course. The chemicals might have evolved independently in each species following their split. Nonetheless, attempts at reconstruction often leave reverse engineering as the best—if not the only—currently available option for reconstruction.

A TAKE ON RECONSTRUCTIONS

Given less than the desirable amount and specificity of evidence, reconstruction can be approached in two ways. One approach is to forgo the exercise and avoid inciting skepticism and criticism among readers. A second approach is to try to develop a plausible scenario with the aim of stimulating further research and interpretation. I have adopted the second approach. That said, it needs to be affirmed strongly that no two individuals writing on the evolution of belief and *divides* would develop similar accounts.

The reconstruction begins with a selective overview of events, contexts, and behavior that might have been critical in the evolution of belief and *divides*. It then turns to the pivotal functions of language, observation learning, emotion, and cognition. The chapter closes with a discussion of the migration of *Homo sapiens* out of Africa and the dispersion of beliefs.

BONES AND TOOLS

Belief and *divides* didn't appear overnight. Multiple requisite conditions and events, some dating back millions of years, presaged their appearance.

There are skulls and bones of human ancestors that are at least 4 million years old.[9] Bones used for tool-assisted consumption of animal tissue may be older than 3 million years.[10] Single-edge cutting tools— "Oldowan choppers"—were in use 2.6 million years ago. Hand axes have a history of at least 1.6 million years. A greater-than-2-million-year history of tool making suggests that the craftsmen of that period had mastered the techniques of tool construction, had an appreciation of the utility of the tools they constructed, and took care to remember where they stored them. A reasonable inference is that capacities for remembering, believing, predicting, and constructing tool kits for practical matters were present and in use.

GENES AND BRAIN CHANGE

The progressive speciation of our own species, as well as other humanoids, is a history written in part in changing genetic-behavior associations. For example, aggressive facial patterns, defensive forelimb movement, and reaching-and-grasping movements appear to be largely "hardwired" in the primate brain and species-universal in form.[11] Such findings hint at their ancient origins as well as their adaptive value through stages of species change.

Genetic changes also altered the physiology and functions of the brain.[12] A possible case in point is the brain chemical dopamine. Well before the first tools were made, it's likely that dopamine influenced the intensity of the emotions of pleasure, reward, and displeasure.[13] Or consider the prefrontal cortex. Today it is integral to cognitive control and harbors capacities to coordinate thought with action reflecting personal goals. No doubt it was present in the distant past, although in a less sophisticated form compared to today.[14] The integration of physiology with brain function is also likely: it is the brain's frontal cortex and midbrain where dopamine largely exerts its influence on emotion and cognition. A similar story applies to the hormone oxytocin. It increases feelings of trust when its levels are elevated.[15] Genetic change leading to conditions that alter its brain levels may have been instrumental in the early moments of extended socialization.

Complementing hardwiring are partially innate capacities that were refined in response to gene-brain-environment interactions. Examples include behaviors and capacities that process environmental and social information using touch, visual, olfactory, and auditory modalities, possibly also associational learning and episodic memory.[16] The refinement of these capacities and behaviors is likely to have correlated with an increase in the size of semipermanent social groups along with improved strategies for mastering novel environments and challenges. Structural changes in the brain would follow these developments.[17]

Then there is novel behavior, that is, behavior that is describable by its form and definable by its uses but that is not species universal in form. This is behavior that is not the direct result of natural selection but that emerges when existing systems or capacities combine in new ways in response to social and physical environment options.[18] The multiple skills required for constructing single-edge cutting tools, engaging in cooperative hunting using coordinated strategies, and constructing living structures are examples.[19] The dates when these behaviors first appeared are unknown. But clearly, when they did appear, our ancestors were poised to exploit their potential. They would continue to do so for thousands of generations to come. Thus it is not surprising that a large percentage of today's human-behavior repertoire is composed of novel behaviors. Driving an automobile in reverse and playing the piano are examples. That such behavior can influence the structure and function of the brain as well as subsequent evolution is suggested by studies showing that mastering reading results in changes in the cortical networks for vision and language.[20]

There is yet another chapter to this history. Genetic, environmental, and experience differences meant that there would be individual differences in personality, emotional makeup, mate and lifestyle preferences, physical skills, food tolerance, and capacities for processing information.[21] Such differences are compatible with the view that multiple and very different beliefs were present in the past, much as they are today. An unavoidable implication of these points is that no two brains were likely to have believed or managed *divides* exactly the same way.

Despite these many contingencies, the net effect for our ancestors was a gradually expanding understanding and mastering of their personal and social worlds and their physical environments. These developments would serve as the matrix for the emergence of culture, increasingly complex social organization, and a continuing diversity of beliefs.

CULTURE

Culture can be defined as the beliefs, social forms, and material traits of a group that comprise a body of knowledge and ritual that is transmitted across generations.[22] There are various estimates for its starting date, although the notion of a starting date is misleading. Culture didn't just appear at a specific moment in time. It evolved gradually over millennia. Estimates vary as to when there is clear evidence of its presence. The date of eight hundred thousand years ago has been suggested.[23] Some scholars would place it farther back in time—this would be consistent with the evidence of the far earlier presence of tools. Others favor a more recent date. The *divide* is indeterminate and influenced by how culture is defined. I will adopt eight hundred thousand, plus or minus two hundred thousand, years as a reasonable estimate: during that period, members of an as-yet-to-be-identified hominoid species, whose artifacts suggest the presence of culture, settled in northern Europe.[24]

Culture is a product of people living together, interacting, and sharing information. Mothers and their offspring were likely the first to do so. The probable next step was the formation of small groups composed of family members and then, eventually, more distant relatives. As group size increased, so would the collective knowledge of groups. Perhaps, too, these events were associated with occasional inbreeding with now-extinct species along with a glimmering of concern about genetic continuity. Current genetic evidence is consistent with the cross-species breeding possibility.[25]

Culture presupposes the presence of a social brain, that is, a brain that can meet the computational requirements of group living. These include pair bonding, processing a diverse array of one's own and others' emotions, such as empathy, anger, and pleasure, and recognizing and tolerating individual differences in goals, belief, and action.[26] The social brain also means collecting and interpreting information using multiple brain channels, including those that process posture and facial expressions and auditory information, such as voice tone and voice signatures. Each of these channels served as a source of evidence that likely fostered various forms of social cooperation, including direct and indirect reciprocity, coalition formation, altruism, sensitivity to fairness among group members, and beliefs about how best to behave.[27] There were also likely effects on the brain. For example, among adult humans, amygdala volume positively correlates with the size and complexity of a person's social network.[28]

An interesting possibility is that groups composed of individuals with similar attributes, such as personality type, food tolerance, and brain levels of the chemicals dopamine and serotonin, sought each other's company and migrated together. This would represent an example of the *founders effect*: a subset of a population with specific genetic characteristics finds itself isolated from other groups and a larger gene pool.[29] At times, isolation appears to be followed by the rapid alteration of attributes. This is one way to read data documenting the genetic, behavioral, and belief differences among groups that, for generations, have been isolated reproductively and socially.

Living together invites a division of labor. Some individuals hunt, some raise children; some lead, and some follow. With increased experience, behavioral options and constraints such as preferences for group size associated with different tasks likely became important.[30] So too for joint decisions among small groups—in effect, a kind of collective intelligence—which often is superior to decisions made by single individuals.[31] But living together also opened the door for social manipulation, such as deception and free-riding.

A social brain further implies that there is a sense of the self, that is, a personal identity that, among other things, permits its owner to engage in social comparisons.[32] This sense was likely associated with beliefs such as judging oneself more attractive, brighter, or more skilled than others as well as the development of *divides* dealing with evidence deemed to justify such judgments.[33] Social comparison also has a history.

Beliefs, probably less complex than those of today, were an essential part of these unfolding developments. Yet even during the moments when culture was in its infancy, it is probable that our ancestors had beliefs about such matters as the advisability of certain social behaviors and relationships, raising offspring, hunting, animal behavior, and the dangers of interacting with other groups.

LANGUAGE

The emergence of culture presupposes the development of a shared means of communication for establishing relationships and facilitating group management.[34] The origins of human language and near-human speech may date back at least two million years, to the Old Stone Age.[35] Likely early uses were among mothers and offspring as well as craftsmen and hunters who shared their techniques. Whatever the details, language appears to have been a critical facilitator in the evolution of our ancestors. Understanding and exploiting environments and carrying out activities such as group organization, migration, and group defense could follow.[36] As with other evolved capacities, language development no doubt underwent thousands of years of refinement punctuated by multiple, small, genetic changes associated with the ability to speak and learn.[37]

If, as some scholars have suggested, humans evolved primarily to act to survive and reproduce and only secondarily to think, there are implications for language development.[38] A preponderance of words and expressions that identified predators and environmental dangers, warned group members of such dangers, and facilitated mating would be expected. Conversely, had humans evolved primarily to think and reason, and only secondarily to act, then thinking and reasoning would be more efficient and responsible for fewer errors than likely was the case in the past and certainly is so today. The lingering effects of this possible early asymmetry may still be with us: we appear to be much better at naming things than reasoning clearly.

An understanding of the evolution of language is far from complete. For example, it is not clear how, after millions of years of evolution, the brain could have the capacities for reading and writing, which began only some 5,400 years ago.[39] The fact that researchers currently are looking at communication among nonhuman species with the aim of obtaining insights about language development among early humans is also a sign of incomplete understanding.[40] Completing these investigations may require decades. Nonetheless, two points invite comment. In the distant past, culture and language bidirectionally influenced each other much as they continue to do today. And language and belief would become inexorably interwoven.

OBSERVATION LEARNING

Very likely the refinement of observation learning accompanied and accelerated the emergence of culture and behavioral change. This capacity also appears to have origins in the remote past: it is observed, for example, among living orangutans.[41] Observing how others behave and accomplish tasks often improves an observer's subsequent performance.[42] Such learning also may be a source of belief transmission.[43] Young children seem to recognize this as they observe and imitate their older siblings and friends. And observing how others do things is a standard feature of today's trade schools. As groups increased in size and knowledge, the "learning yield" from observation likely increased significantly.

What is learned by observation is direct evidence. Hearing what others have observed or do is indirect evidence. Compared to direct evidence, indirect evidence is more vulnerable to misinterpretation under conditions of limited language and conceptual capacities. Thus, early on, the number of beliefs built from direct evidence probably exceeded those based on indirect evidence.

It is also worth considering the possibility that some features of human behavior were learned from observing other species. Chimpanzees are known to use sticks to poke into anthills to collect ants for eating. They crack nuts with rocks. They organize their groups hierarchically and defend their territories. Wild dogs and hyenas hunt in packs just as early hunters are thought to have done. Birds store food for the winter, and some species of monkeys wash their food prior to its consumption.

EMOTION

At this moment in time, we are a very cognitively biased species. We place great value in how we reason and in what we believe. We defend our beliefs and reasoning often with surprising intensity. We probably do this more often than we defend our emotions. Thus, on first pass, it may seem improbable that among our ancestors, emotion had a privileged status in the brain relative to cognition. What was felt was given greater consideration than what was thought.

Chimpanzees can serve as a model for this early emotion-cognition asymmetry. They have an emotion-based vocabulary dealing with social interaction.[44] They often exhibit anger, which is associated with threats and physical fights. They have emotional responses to dangers posed by other groups of chimpanzees and predators. A reasonable guess is that, following the split of chimpanzees and our ancestors from a common source, early human ancestors had a similar emotion-based vocabulary and information processing system. A related guess is that beliefs lacked the complexity, and *divides* lacked the precision they would eventually achieve with the emergence of more-refined cognitive capacities and language.

Among these distant relatives, specific behaviors were likely associated with emotions such as pleasure, displeasure, pain, and boredom.[45] Eating X was pleasurable and reduced the unpleasantness of hunger. Drinking water when thirsty worked the same way. Visceral change and satisfaction followed. Participating in sex was pleasing. An unanticipated encounter with a large threatening animal was frightening and to be avoided in the future.

It is important to note that many emotions have origins in locations other than the brain, although the brain processes them using a complex network of nerves radiating throughout the body. For example, indigestion begins in the gastrointestinal system. Pain may begin in the toe or in the elbow. Thirst often seems to begin in the mouth, but its origins are elsewhere. Pleasure too can begin at multiple locations, as when one sits down to relax from a day full of tension or overwork just as it can seemingly begin in the back while receiving a gentle massage. That actions associated with pleasurable feelings were repeated more than those that were unpleasant is likely. Repetitive actions suggest that memory was in place and associated with decisions. Such repetitions can be viewed as early forms of belief in the sense that belief is often associated with prediction and action. Studies of nonhuman primates are consistent with this view. They show that reward history predicts foraging behavior and that systems in the parietal cortex process the relative value of competing actions, that is, they anticipate reward versus less-reward outcomes.[46]

In time, emotion would find a similar place among the positive and negative physiological effects of social interactions. These are known to affect the activity of specific brain areas and brain physiology.[47] For example, a serious threat by another has a direct impact on a recipient's physiology, such as raising the recipient's level of adrenalin; conversely, a warm smile and positive greetings do the opposite.

To state the preceding points another way, pleasant and unpleasant emotions served as primitive, largely noncognitive body-experienced indicators of the consequences of action or inaction. They are types of direct evidence and they are believable. As a result, emotion likely established its place as a key factor associated with future action.[48] Further, the ratio of the emotions of pleasure and displeasure may have served as a subjective measure of well-being.[49] In turn, the measure may have provided a hint about the "when" of migration: when the sense of displeasure exceeded that of pleasure, it was time to consider moving on.

To bring these points into focus as they apply today, consider how many things we do based on similar emotion-action associations and the propitious predictions to which they lead. For many, a cup of coffee in the morning is associated with the pleasurable feelings of being awake and energetic. It seems to work every morning, even though one may possess literally no knowledge of the chemistry of caffeine and its effect on the brain. Or take lovemaking, which can be delightful without a detailed understanding of what is going on in the body and brain. The same point applies to work preferences. Some people are more energetic and creative in the morning. For others, it's the afternoon. Because it's likely to be that way tomorrow and next week, people arrange their days accordingly.

THE COGNITIVE EXPLOSION

From the beginning of what has been characterized as *Homo sapiens'* cognitive explosion—circa forty-five thousand years ago—the fundamental nuts and bolts for complex cognition, belief, and *divides* were present.

How cognitive systems evolved is still an unanswered question.[50] But clearly the precursors of these systems were undergoing change and refinement long before forty-five thousand years ago.[51] In time, they would overshadow the primacy of emotion-belief and emotion-action connections—recall that Descartes said, "I think, therefore I am," not, "I feel, therefore I am." There are steps preceding the explosion that seem likely, such as the brain developing capacities to imagine and rehearse before speaking.[52] The capacity to categorize information, such as identifying dangerous and non-dangerous animals or plants, is also likely. Developing inferences about the intentions of others by evaluating their actions is yet another example. That nonhuman primates have many of these capacities hints again at their possible ancient history. As the explosion unfolded, it is likely that our ancestors could plan well ahead in time, think and speak in long and nested sentences, evaluate the utility of behavior, and believe in ways similar to people today.

The cognitive explosion was the outcome of thousands of minor steps in evolution leading to a highly sophisticated brain that is efficient in a variety of functions, such as remembering, planning, decision making, believing, creating and responding to music, and, at times, estimating risk.[53] Most important for this book, these functions often seem to operate independently and compete with each other. That this is the case is suggested by the frequency with which our emotion and cognitive systems are responsible for beliefs that are associated with actions that backfire and *divides* that misinform. Emotion often overrides reason and cognition and vice versa—they are intertwined.[54]

MIGRATION OUT OF AFRICA

The prevailing view among today's scholars of human evolution is that anatomically modern humans evolved somewhere between 200,000 and 160,000 years ago.[55] At some time, circa 100,000 to 60,000 years ago, they began migrating out of Africa, although what route or routes they took is currently a topic of debate.[56] With migration, social and environmental knowledge would increase and culture would notch up in complexity.

Dispersion out of Africa to different environments meant a parallel dispersion of beliefs and *divides*. Different subgroups of *Homo sapiens* were changing in response to different demographic challenges. Genetic changes followed.[57] Beliefs and *divides* would not necessarily get simpler: the greater the variety of experience and knowledge, the greater the potential for different beliefs and *divides*.

Despite the many mishaps that befell individuals and groups, our ancestors would eventually occupy all the major ecosystems and landmasses of the world.[58] The refinement of capacities dealing with the manufacture of material goods continued in parallel. For example, changes in tool manufacturing track over the centuries from millions of years ago to forty-five thousand years ago, at which time, there occurred an exponential increase in their complexity.[59] A rapid increase in computational capacities—essentially, "intelligence"—may explain this surge in cultural and creative complexity, although rapid changes are rare in recent human evolution.[60] (An alternative interpretation for the explosion is introduced in chapters yet to come.) Through these changes, natural selection continued as indicated by recent genetic studies that suggest that humans have undergone detectable genetic change since the end of the last ice age some ten thousand years ago. Indeed, even today there may be continuing genetic divergence among remotely located groups due to the selection effects of different physical and social environments.[61]

This brief and speculative review has been aimed at identifying antecedent features of brain, behavior, and contexts associated with the evolution of belief and *divides*. Our ancestors evolved due to natural selection, encountering novel environments, and developing culture. Capacities to connect specific actions with emotions of pleasure and displeasure appear to have been active before the refinement of capacities we normally refer to as cognitive. With time, this relationship has reversed. An increase in computational capacities may explain the increase in the complexity of culture and creativity that occurred some forty-five thousand years ago.

12
ENTER IMAGININGS, BELIEFS, UNCERTAINTY, AND AMBIGUITY

Four days after sending a draft of this chapter, Greg's e-mail arrived.

Hi Michael,

The chapter is fascinating but, as you no doubt suspect, I have questions. But first, some of my experiences, which map to your text. I agree that imaginings and beliefs often "pop" into one's head. Moreover, when I try to figure out what might have caused them I usually fail. I distinguish such moments from what I would call associations. If in walking past a café I imagine that it would be delightful to be sitting there having a conversation with you, that's an association—I wouldn't have imagined it unless I passed the café. Such imaginings contrast with head-popping when there is no obvious stimulus—for example, the other day I was walking through the countryside, enjoying the scenery, and out of nowhere I imagined a fix to my 1937 Chevrolet, which is stored in a garage in Boston. Much the same is true with beliefs. They too pop into my head and often at odd times. Usually these are not beliefs that I develop after carefully attending to evidence or, in many instances, ever thought of before.

It's your distinction between ambiguity and uncertainty that troubles me most. I have read and reread what you've said. I'm still uneasy. They seem to overlap more than they are separate. Could they be examples of false categories?

Still, as said, overall a fascinating and very informative chapter —so go with it.

Regards,

Greg

EVOLVED BRAIN SYSTEMS

Belief and *divides* are products of brain systems that have been undergoing change and refinement for millions of years. A similar history is likely for *imagining*, *mirroring* (replicating the brain activities of others), *Theory of mind* (reading others' brain states), *attributing* (assigning attributes to persons and events), and *triggering* (events initiate specific brain states). Alone or in combinations, they are in part responsible for beliefs and *divides*. Imaginings, belief, ambiguity, and uncertainty are the topics of this chapter. Mirroring, Theory of mind, attributing, and triggering are discussed in chapters that follow.

IMAGINING

When one of my sons was five, we spent a day at the local zoo. He found it delightful. The chimpanzees entertained us with their tricks. Parrots gave us advice. Giraffes were running back and forth for no obvious reason. The lion looked directly at my son and growled; after a moment of crying and hiding behind me, he recovered. We visited the aquarium, fed ducks and goats, and went for a ride on a pony. On the drive home, he asked if he could live in the zoo with the animals.

Once we were home, he went to his room and soon emerged with a picture of an animal with the body of an elephant, the head of a giraffe, fins of a fish, and feet of a duck.

"Dad, look! It's my new friend."

"Let's see. You're friend, does it have a name?" I asked.

"Boo, it's Boo."

"And what is it?"

"It's my friend, Boo," he repeated.

"Did we see an animal that looked like that at the zoo?"

"No, but he's there."

Imagining is an act of forming a mental image of what is not actually present or as experiencing something never before experienced. Described another way, it is an act of the brain associated with awareness of what is not currently sensed.[1]

Scholarly writings distinguish between imagining and belief. What is imagined is not thought to initiate action, while belief often is assumed to do so.[2] On close evaluation, this distinction seldom holds up. Imaginings resemble beliefs in several ways. We know of them because we experience them. We remember them. They are products of unperceived brain systems that are active prior to the time we become aware of them.[3] Their representations are assumed to have energy, mass, and location in the brain. They may be associated with action; for example, I may imagine that cleaning my disorganized garage is necessary, following which, I find a broom and a dustpan and go to work. There may be no limits to such creativity.

Imaginings differ in their content. Usually they include features of the familiar, such as envisioning the desirability of a cold drink of water when one is thirsty but no water is available or imagining a novel animal after a day at the zoo.

It is uncertain at what time during the past imagining evolved. Hints come from studies of nonhuman primates. They are known to possess problem-solving capacities and exercise strategy choice in response to both novel experimental challenges and *fictive* outcomes (possible rewards or punishments that have not yet been experienced).[4] Other studies suggest that they possess the capacity to causally connect events that occur close together in time.[5] Thus they may possess information-processing systems similar to those of humans, which means that the rudiments of these systems likely extend far back in the past.

Early imaginings might have occurred in response to natural events, such as earthquakes, floods, or eclipses. They are strong candidates for eliciting attention and leading to cause-and-effect connections among those experiencing them. Feeling the earth quake could lead to an imagined cause, such as the earth being alive. An eclipse could be conceived of as the work of an unknown entity altering solar events. Or it could work the other way; an earthquake or an eclipse could be conceived of as a cause.

Subsequent events, such as an outbreak of sickness, the birth of triplets, or unusual behavior by a familiar animal, might be the imagined effects.

Biological differences in emotional and cognitive states are other possible sources as well as causes of different imaginings.[6] For example, people with normally high levels of brain dopamine activity are more likely to find significance in correlations and more frequently make cause-and-effect connections compared to those with low activity levels.[7] High levels of the brain chemical serotonin correlate with an optimistic outlook and possibly an above average tolerance of individual differences and day-to-day disappointments.[8] Sensation seeking and impulsivity are also possibilities. Recent studies have linked these behaviors to atypical genetic profiles and above average brain dopamine activity.[9] These behaviors may be associated with imaginings such as contemplating highly risky acts as exciting or conceiving of causal relationships, which vary from normative views of how the world works. Or, consider studies that show that when two people are watching the same event, such as a movie, their fMRI profiles are similar. However, when the movie is over and they discuss their interpretations, they often differ significantly.[10] These studies not only suggest that there may be no species-characteristic way of interpreting experience but rather that individual differences in interpretations may be the rule, not the exception—so too possibly with imaginings.

Then there are mind-altering drugs. It is not known when our ancestors first experimented with these drugs. There is, however, cave art dating back at least seventy thousand years depicting social gatherings, hunting, and supernatural entities. Much of this art is thought to be the work of artists influenced by drugs that are associated with novel states of awareness.[11] This possibility is consistent with what is known about brain-altering drugs, such as LSD (lysergic acid diethylamide), which leads to a distorted sense of time, the presence of radiant colors, and visualizing crawling geometric patterns.[12]

Dreams are yet another possibility. Often they consist of parts of experiences reconfigured in novel ways similar to moments when the brain is wandering.

Differences in memory capacities also need to be considered. Many of the relatively stable features of memory storage and recall among today's humans are likely more refined and less varied than in the past.[13]

Then there is a host of imaginings that remain to be explained, such as Atlantis: fact, fiction, or a mixture?

Whatever the details, the point to stress is that imaginings were experienced and interpreted. In turn, a gradually expanding library developed composed of imaginings and their interpretations. Some would become beliefs, predict future events, and become associated with behavior. Others would not.

To return to the daily lives of our ancestors and their early moments of migration, lack of knowledge and uncertainty about what lay over the horizon likely were influential and disturbing factors vis-à-vis decision making. Early migrants had only themselves as information sources—no maps, travel guides, or access to Google Earth. Thus, a reasonable guess is that they were guided by imaginings such as "there may be desirable land over the horizon," "fewer predators may live in the next valley," and "more food may be available across the river"—in effect, the integration of bits and pieces of experience as they might apply to locations about which specific information was lacking. For such imaginings, past experiences would have served as a constraint: individuals living in central Africa and lacking familiarity with large bodies of water were unlikely to imagine trips across the sea, while those living at the seashore might have done so.

TWO TYPES OF MEMORY

Early forms of two types of memory likely affected both imaginings and belief.[14] Episodic memory, or the recall of prior experiences, is one type.[15] Working memory is another type: it refers to the ability to hold information in memory and manipulate it.[16]

With episodic memory, past experiences could be recalled and behavior that worked successfully yesterday could be repeated today—the tool kits that have been mentioned. Working memory could facilitate the reorganization of what was being recalled and serve as the basis for imagining novel strategies, such as cooking meat differently to alter its taste and digestibility or changing hunting strategies to improve their success.[17] These changes likely built primarily on the recall of schemata—essentially, direct evidence of experiences—which retain features of events that are lost in the verbal recall of an event and are absent with secondary evidence.[18]

Episodic and working memory also were likely contributors to early *divides*. But episodic memory can be very inexact, something psychoanalysts and laboratory studies of memory established decades ago. Should this point be doubted, try recalling the details of a movie you viewed a year or two ago. Many of its details will be remembered incorrectly if at all. Episodic recall thus may be accurate or not. The catch is that it is often difficult to distinguish between the two possibilities. Add to these points the already-mentioned *divide*-reduction effects of data distortions: disconfirming cases are ignored, and selective remembering leads to recalling more confirming than disconfirming experiences.[19] In effect, imaginings may find their origins in inaccurate memory and information manipulation.

The preceding points add up this way. Among our early ancestors, differences in information-processing capacities were likely greater than they are today. Further, early imaginings were more likely to have become beliefs than is the case today—this is because of the relative absence of experiences and secondary evidence that might disconfirm what was imagined. Without such information, there would be an enhanced bias favoring the narrowing of *divides*.

BELIEF

Two weeks after our visit to the zoo, my son wanted to return. I was busy that weekend and told him that we would have to delay the trip until the following week.

"But I want to go see Boo; he misses me," was his reply.

"Are you sure we saw Boo when we were at the zoo?" I asked.

"Yes. He was there. Dad, I want to visit him; he will miss me."

I changed my plans and we went to the zoo. No Boo was found. On the way home, my son offered why: "He left the zoo and he is coming to live with us."

Six months passed before Boo was forgotten.

Following the emergence of imaginings, a likely next step was that some imaginings became beliefs and were associated with predictions. As mentioned, this may have occurred in response to natural occurrences, such as an eclipse. But beliefs were also likely to have developed in association with less dramatic events. Eating an unfamiliar root or an unfamiliar part of an animal, if followed by serious indigestion, would lead to connecting the two events and the belief that one caused the other. So too for excessive fatigue following a hard day's work or the scratch of a thorn followed by a painful infection. Such connections are often highly predictive, narrow *divides*, and are associated with action.

A probable accompanying step was belief generalization. For example, it might have been imagined that hunger would occur while climbing a mountain. Before climbing a mountain for the first time, there was no certainty that it would. The *divide* was indeterminate. But once, or maybe twice, or even thrice, when a climber experienced hunger while climbing, the uncertainty would disappear and be replaced by belief and prediction of future behavior: what happened today will take place in a similar fashion tomorrow and next week. What was once an indeterminate *divide* narrowed or disappeared. In turn, beliefs associated with predictions could be generalized: hunger will occur while climbing mountains north, south, east, and west. Our current understanding of the brain hints at how such generalization might have taken place: unperceived imaginary rehearsal, which improves the precision of predicting outcomes of future actions, is likely to have become progressively refined and reduced *divide* distances.[20]

Beliefs dealing with cause-and-effect interpretations are consistent with ethnographic reports from so-called primitive societies. Events ranging from the mundane to the spectacular are explained by both likely and unlikely causes or, at times, vice versa.[21] These creative imaginings and beliefs offer insights into many of the brain's repeated inventions over much of history such as gods, mysterious forces in the universe, and animals and inanimate objects with personalities and other humanlike attributes. Nor have imaginings and beliefs changed much compared to the distant past. One might expect that over time a body of beliefs would build in which those that consistently predict outcomes would trump those that don't. Poor and inconsistent predictors of future outcomes or events would disappear. But among twenty-first-century humans, imaginings and beliefs dealing with gods, afterlives, conspiracies, the economy, the private lives of others, overnight solutions to complex problems, and unrealistic strategies for personality change are everywhere apparent.[22] We see what we believe.

AMBIGUITY AND UNCERTAINTY

Greg's e-mail had its effect. His concern about the overlap of ambiguity and uncertainty kept spinning around in my head. Things were still spinning weeks later when I arrived in Mexico to visit my sister and her husband. She and her husband, an Italian and former member of the Alfa Romeo race team, lived on a ranch outside Taxco. One afternoon, during a quiet moment, I picked a book from their library. It identified a dozen lost cities of the world, cities either that myth said had existed or for which there was evidence of their existence but they couldn't be found.

Rio Bec B was one of the cities. It was first identified near the border of Quintana Rue (Mexico) and Guatemala in the early years of the twentieth century. Subsequent expeditions to locate it had failed to do so. At the time, it was believed to be one of the last major cities built by the Maya. Its potential importance stemmed from the possibility that it might help explain the implosion of the Maya civilization.

I suggested to my brother-in-law that we should try and find it. The suggestion clicked. The idea of searching for a lost city excited him. Eight months later, our search began.

The area in which we searched was dense jungle. Aside from the already-mentioned jaguars, available reports noted that the area was inhabited by highly aggressive ants and a variety of poisonous snakes, including the bushmaster, the fer-de-lance, the coral snake, and the tropical rattler.

This story is not about finding Rio Bec B, however. We didn't find it, despite spending two weeks tracking about the jungle. Rather it's about moments of the search, poisonous snakes, imaginings, uncertainty, and ambiguity.

One morning, weary of living in the semidarkness of the forest and hoping to get a dose of sunlight close to our camp, we cleared an area of its bushes and small trees so that rays from the sun could reach the ground. We then departed and continued our search. Two hours later, we returned to the camp. There in the sunlight were six snakes, all poisonous.

We had imagined that snakes lived in the area, but we were uncertain if they did. We had carefully watched for them. Yet until that moment in the sun, none had been seen. Once they were observed, our uncertainty about their presence disappeared only to be replaced by a new uncertainty: How dangerous was it to continue our travels through the forest? Our excursions in the jungle continued for another ten days, during which we were hypervigilant about snakes. Yet we never saw one again, except for one day when everyone was away from camp for several hours: snakes again had returned to the opening.

Ambiguity and uncertainty are states of awareness with physiologic, brain-system, and behavior features. They are unlikely to be brain systems, per se, but rather brain states that occur under certain conditions. Ambiguity refers to doubt about explanations of states or events that may be inexplicable or subject to several interpretations. Strange and unexpected behavior by close friends and trying to fathom the causes of complex social events, such as unpredicted population uprisings, are examples. Uncertainty refers to the absence of a clearly defined state or event. The suggestive appearance of life-forms at dusk often qualifies, as does the possibility of an afterlife for many people. Even in the physical world, ambiguity and uncertainty can't be avoided: the world we perceive doesn't reliably conform to the popular conception of causality, which assumes perfectly predictable cause-and-outcome relations.[23] On the other hand, these states often can be avoided when one encounters a situation for which one has a ready and reliable tool kit with which to respond.

Ambiguity and uncertainty are not necessarily due to external stimuli. The brain's information-processing systems may be their source. Illusory correlations and other types of information distortions have been mentioned. Or they may follow on the footsteps of a belief: for example, believing that a god exists can quickly lead to ambiguous or uncertain moments, such as wondering how best to please the god, whether the god is a friend or an enemy, or whether the god prefers some people to others. Similar situations apply to very practical situations: believing that one's boss is the key to one's future job promotion can lead to concerns about how to act, what to wear, what friends to make, and so forth. Because such situations are often aversive, a frequent response is to create an elaborate system of imaginings and beliefs to counter their undesirability.

A feature of both ambiguity and uncertainty is that they increase computational requirements. Contemplating the many possibilities of a potentially life-threatening operation is an example. The ambiguity associated with the choice of a doctor and the uncertainty of the outcome of the operation often won't go away.

It is likely that the early-human brain of the remote past didn't seek ambiguity and uncertainty any more than most people do today. Most of the time, we prefer certainty, clarity, and predictability. Uncertainty and ambiguity are associated with aversive changes in specific brain chemicals that have undesirable effects.[24] Further, fMRI-identified signatures in the brain are known to be present and different types of ambiguity activate over two dozen areas in the brain. For example, studies show that the brain's emotion-management center, the amygdala, is critical in initiating a sense of caution toward engaging in behavior in which the outcome is uncertain. The *divide* is indeterminate. Yet people frequently disregard the risk and act. Why?

A likely answer is that the brain isn't sitting by idly. It has its own set of remedies for managing undesired physiologic effects. Recall that the brain has many independent agendas, and they are carried out without the intention or awareness of effort on the part of their owners. One of its favorite remedies is to reduce unpleasant states by developing beliefs and narrowing *divides*. To do so has proven to be an efficient energy saver. This happens when a new mother discards her worries and uncertainty about her infant's "normality" and comes to believe that her child will grow up as a healthy, bright, and responsible human being. It occurs during moments of serious illness when one decides that one's doctor is the best there is and that his recommended intervention will work as one hopes. Such beliefs lead to desirable physiological change.

13

THEORY OF MIND, MIRRORING, AND ATTRIBUTION

Scholars in disciplines ranging from philosophy to psychology agree that believing that we can truly know the brain states of others is an illusion.[1] The best we can do is make guesses, draw inferences, and engage in post hoc constructions. This holds even when others "reveal all" about themselves. Yet often we are convinced otherwise. We *do* believe we know other's mental states. Actions associated with such beliefs often follow. When we choose friends, select spouses, hire or fire employees, or send individuals to jail, our beliefs about their brain states are involved. When people behave as we predict, our views seem confirmed. When the opposite occurs, it's time to revise our views, although the frequency with which this happens remains unknown.

The preceding points might suggest that it's a waste of time to try and clarify what happens in our own brain and in the brains of others. This is not the approach adopted here. Some properties of the brain are known and documented. Others can be inferred with reasonable certainty.

This chapter focuses on three postulated brain systems: (1) *Theory of mind*, which deals with how individuals surmise the brain states of themselves and others, (2) *mirroring*, which explains the effects of other's behavior on the brains of observers, and (3) *attribution*, which involves assigning attributes to oneself and to others and explaining behavior, natural events, and physical systems.

Again, I offer some words of caution. With the exception of mirroring, which has a scientific basis, Theory of mind and attributing are concepts that are difficult to discretely separate let alone precisely measure. They often seem to blend together, which may be due to the sharing of parts of a common operational anatomy in the brain.[2] But we do experience moments of awareness that can be interpreted as a result of these systems. It is these moments that I will use as points of departure.

But first, some comments about computations.

COMPUTATIONS

The two preceding chapters contained discussions of the brain's computational requirements—essentially, the number of bits of information that need to be processed per unit of time—for early modern humans living in small hunter-gatherer groups approximately two hundred thousand years ago.[3] Many of the systems that affect these requirements have been mentioned, such as language and the access, storing, and manipulation of information.

In chapter 12, it was noted that some scholars have suggested that the upswing in sociocultural and inventive complexity, which occurred approximately forty-five thousand years ago, was the result of an increase in computational capacities. As this interpretation goes, our ancestors became brighter and their computational capacities increased. This may have happened. It's a plausible hypothesis.

There are alternative possibilities however. For example the refinement of already-requisite systems, not an increase in computational capacities, might explain the upswing. Armed with an ever-enlarging number of reliable tool kits dealing with social interactions and for mastering the animal and physical environments, the brain is likely to have improved its efficiency. Already-present computational capacities may have been sufficient for this improvement. Hence the possibility of increased creative complexity without an increase in intelligence or energy costs to the brain. Theory of mind, mirroring, and attribution possibly fit this scenario.

THEORY OF MIND

A discussion with a student.

> *Author:* "And what makes you think that Professor Q doesn't like your doctoral thesis?"

Student: "He hasn't said that exactly. But I sense it. There are indications. He talks with other students more than with me. Most of them are working on topics he likes. I'm not."

Author: "Let's be more specific."

Student: "He has published books on how people—wealthy people—in the South exploited the slaves. That's what he believes. He is right, but it isn't the whole story. My thesis is about the other side of the story."

Author: "And the other side of the story is?"

Student: "It's about how many Southern families provided the slaves with a safe place to live, work, and to develop their own interests. That happened too. Many became members of the families. They raised the family's children. They gave advice about how to run the plantations and the advice was followed."

Author: "OK, but how does it follow that he doesn't like your thesis?"

Student: "He never says whether he thinks I'm right or wrong, as he does with other students. He just says things like, 'Are you convinced about your sources or have you looked at other explanations?' We seem to get nowhere."

Author: "Are you afraid that he might reject your thesis?"

Student: "It's a serious concern. That's why I came to talk with you."

Theory of mind refers to the capacity to surmise mind states. "Mind reading" is the usual term when we surmise the mind states of others. "Self-reflection" is the usual term when we focus on ourselves. The capacity to mind read is thought to be mostly innate and subject to refinement. Forty percent of adults are reported to believe they can read the minds of other people.[4]

Given the monistic view of the brain adopted here and the many often contradictory uses of the word *mind*, the time is ripe for a change in terminology: *Theory of brain*, rather than Theory of mind, and *brain reading*, rather than mind reading, will be used from here on.

Descriptively, what is involved is simple in principle and familiar in practice: one tries to put oneself in the shoes of others. One listens to what others say and observes their behaviors. Direct and indirect evidence is assimilated. Memories are created. Gradually representations of the brain states of others are developed and may be experienced in awareness.[5] Representations may change with new information or a change in context. Similar events occur during self-reflection.

Brain reading goes on all the time. How many times each day? The answer is hundreds. Statements such as, "He is depressed," "The president is planning to oppose the budget," "The Council members ignored X, Y, and Z when they reduced taxes," "He avoids me because we compete," and "She plans to seduce him," stem in part from brain reading. Although at times we struggle to understand others, the majority of time, our sense that we are reading their mental states is automatic and effortless. Little notice is taken of the process.

In the academic literature, brain reading builds on a number of assumptions and preconditions, such as one must be aware of the products of one's brain before recognizing those of other brains and others have brains by analogy to one's own brain.[6] Preconditions include introspection, cognitive and emotional development, and the ability to infer from evidence and to imitate others.[7] These are far from primitive brain operations. Indeed, if the assumptions and preconditions accurately depict what is required to brain read, a highly sophisticated brain is at work.

That we often believe that we know ourselves and others is understandable. We sense we know our own brain states and the causes of our actions. Others also believe they know themselves. We share what we believe with others, and they do likewise. Moreover it is often easy to surmise what is going on in the brains of others: a mother observing her child snitching a cookie from cookie jar is unlikely to misinterpret events taking place in her child's brain.

Research findings are consistent with the view that children are endowed with capacities to brain read.[8] Findings also suggest that upbringing environments significantly influence the refinement of these capacities. Studies conducted decades ago by Rene Spitz and Harry Harlow offer striking examples. Spitz studied the developmental consequences to children deprived of normal maternal care.[9] Harlow separated infant monkeys from their mothers and then evaluated their maternal and social behavior as adults when they were reunited with members of their species.[10] In both studies the effects were similar: socially deprived infants failed to develop into normal adults and exhibited major deficits in the ability to relate to members of their own species.

Other studies suggest that brain reading is compromised in certain mental disorders and following certain brain injuries.[11] Compromise also seems to be the case when interacting with persons suffering from schizophrenia: what they say and how they behave are unpredictable and difficult to interpret. Whether nonhuman primates can brain read is a controversial topic, although there is suggestive evidence.[12] Studies using fMRI designed to identify areas of the brain that are activated during brain reading remain difficult to interpret because of the many areas of the brain that respond to the wide range of experimental challenges used in such studies.[13] Given the networked structure and the often multiple functions of specific brain areas, these findings are not unexpected.

Brain reading has clear implications for beliefs and *divides*. It describes one of the many processes contributing to their formation and management. A generous guess is that our representations of others' brain states are approximately accurate 30 percent of the time, although I know of no evidence to support this guess. Whatever the percentage, readings can quickly become beliefs. As to *divides*, narrowing occurs in almost all instances. It occurs even when two or more brain states are considered such as, "S is either depressed or in pain." In effect, S's possible brain states have been narrowed to two from many, many possibilities.

Views about the brain states of others can change. A personal experience provides an amusing example.

Years ago, while studying monkeys in a foreign country, I was asked by the principal of the local school to give a talk about the purpose of our research. The audience would be members of the senior class. I accepted the invitation.

Developing a view of the students' beliefs seemed easy. I had met some of them. They had asked about the research and were curious about what we were trying to discover. These interactions served as direct evidence from which I inferred that they shared my view that monkeys and humans derive from a common ancestor. The view soon became a belief with a narrow *divide*.

That I looked forward to the talk was clear. Considerable time was spent in its organization and rehearsal.

I arrived at the school. Students filed into the classroom and sat quietly. I was introduced.

The talk began with a discussion about the members of the research group. It then turned to evolution and the idea that humans and the monkeys living in the hills nearby shared a common ancestor. I drew a tree of life on the blackboard and pointed out some of the critical points of hominid evolution over the ensuing millions of years. The talk finished with examples of how humans and the monkeys share many anatomical, physiological, and behavioral features.

As the talk proceeded, I sensed that the students were unreceptive, although no one spoke up. Both the students and the teachers were attentive. No one went to sleep, yawned, or left the classroom. The talk ended when I asked if there were any questions.

An eerie two minutes of silence passed as I stood in front of the audience anticipating questions. There were none.

The silence was finally broken when the principal said: "Thank you very much. We don't believe anything you've said."

The teacher and students departed from the classroom.

The challenge had failed. I hadn't detected that I would be talking to a group in which literally all of its members were religious fundamentalists. My belief was discarded.

As wrong as our brain readings can be, is there an alternative option for navigating the social environment? A possible 30 percent accuracy is certainly better than a lower percentage. Further, an inescapable consequence of group living is that decisions need to be made and actions need to be taken. Many are contingent on reading the brain states of others.[14] For these reasons, the brain states of others are of equal importance to our own.

To reverse engineer some of these points to our ancestors of two hundred thousand years ago, several things seem likely. First, the capacity to surmise the brain states of others was already present. The child of those years wasn't born with a brain that was a "blank slate"—essentially without partially innate information-processing and behavioral predispositions.[15] Rather, much as is the case today, children were born with systems to process, use, and manipulate information dealing with social relationships and for navigating their physical worlds.[16] Second, views resulting from brain reading often became beliefs. Third, brain reading meant that some group members would be preferred to others. Such preferences would favor the development of small groups composed of members with positive assessments of each other. The opposite is also likely: unshared preferences were a factor when individuals disliked or mistrusted one another.

MIRRORING

A friend describes an incident.

> *Friend:* "It made my stomach turn. There he was, an extremely accomplished poet reading his poems to an appreciative audience. Then a heckler began interrupting with crude comments about his work. I was disgusted and mad. I could see his response in his face and body. They tightened up. He looked away. Then there was a look of anger. I was in the first row. It was as if I could feel what he was feeling."
>
> *Author:* "And then what happened?"
>
> *Friend:* "He continued to read and not respond. Then the audience began to yell at the heckler and the reading stopped. The heckler went quiet and we assumed he would shut up. The reading started again, and again the heckler started yelling."
>
> *Author:* "And?"
>
> *Friend:* "Mad as I was, what overwhelmed me were my feelings. It was strange, almost as if I were him, in his shoes, so to speak, the target of the attack. I hurt and I began fidgeting. Then I felt like shouting. Then some university police officers arrived and escorted the heckler away from the reading. The reading continued, but I remained agitated 'til late that evening."

My friend was describing mirroring, which works as follows. Person A observes person B's behavior. Neurons in person A's brain that are the same as those responsible for the same behavior in the brain of person B are activated. In effect, a kind of "copycat" response by the neural cells of an observer in response to what is observed—hence the term *mirror neurons*.[17]

As with brain reading, mirroring appears to be a partially innate capacity that, when refined, operates without awareness or perceived intent on the part of an observer. The process is consistent with the view that the brain is responsive and often controlled by environmental information—the extended brain hypothesis mentioned earlier. (Mirroring is a likely factor in the success of theatre and movies, which can affect the brains of audiences much as do the face-to-face experiences of daily life.)

Although there are those who are skeptical of the phenomenon,[18] research findings strongly suggests that both humans and nonhuman primates possess the capacity. In the cortex of monkeys, single-electrode recordings reveal that mirror neurons fire when a monkey performs a specific action or when the monkey views another agent performing the same action.[19] For monkeys, the other agent can be another monkey or a human being. Studies using fMRI reveal that the human brain regions that contain mirror neurons are active when a person views another's goal-directed action or emotional state.[20] More complex behaviors, such as subtle forms of social rejection, have led to similar findings and interpretations.[21] As with brain reading, studies of mirroring reveal that many parts of the brain respond to what is observed. This is to be expected.[22]

No one has suggested that mirroring is 100 percent accurate. Nor is it assumed to work the same way with everyone.[23] Childhood experiences may sensitize the brain to particular behavior by others. Further, the process is not easily unlearned or constrained. Exercising constraint is one of the tasks facing young medical personnel who must *learn* to remain professionally focused when viewing individuals who have been severely damaged physically or who are in great emotional distress.

An obvious but as yet unanswered question is whether mirroring contributes to belief. It has been suggested that it does:[24] for example, it is easy to imagine that the mirror-neuron activity of a person observing another person in pain initiates events leading to the belief that the observed person is in pain. As with brain reading, such connections initially take place outside awareness.

Are there similarities between brain reading and mirroring? Yes. Both involve the processing of external information. There are also differences. The information processed by mirror neurons may be limited to moments during which observations occur. In principle brain reading can occur separately from observations as during moments in which information that has already been acquired is developed into representations. There are also potential downsides to both capacities. For example, mirroring and brain reading may enhance the effectiveness of deception: neurons responding to a behavior by others but not detecting its deceptive intent may invite erroneous responses and beliefs.

What of *divides*? Mirroring should narrow them for observers. Exceptions may occur however. For example, widening might take place if a person who is deemed to be a deceiver is also in pain. This could initiate a competition between the view of the person and mirroring, leading to *divide* widening—in effect, an unperceived variant of cognitive dissonance where a person is dealing with conflicting beliefs.[25] Is the person being observed really in pain or deceiving or both?

Backward extrapolation suggests that mirroring is an ancient trait, perhaps millions of years old. This is a strong inference from studies revealing its presence among nonhuman primates. As with the other systems, mirroring no doubt went through multiple stages of refinement. Because it replicates activity in specific areas of the brains of observers, it likely fostered sensitivity to others during the early moments of group development. This would have increased the probability of social cohesion, such as assisting others in distress. Or at moments when the joy of others is contagious, one may find oneself giggling.

ATTRIBUTION

As a youth, I spent a month each summer visiting my mother's parents in Mojave, California. Mojave was then a small town of no more than a thousand people. After several visits, I came to know the few boys my age who lived there year around. So when I arrived for my visit at age nine, my friendships from previous summers were easily renewed. It was with my friends that I spent much of my time.

One afternoon they took me to see "the haunted house."

It had its own story. The house had belonged to a prospector who had disappeared years earlier. No one knew why. A mailbox in front of the house was always empty. Coyotes and stray animals were often observed on the grounds.

What I saw was consistent with what I was told. An uprooted tree was on its side in the yard. Vines covered most of the house and roof. Pieces of siding had fallen to the ground. Newspapers, broken sawhorses, worn out tires, and garbage cluttered the property. A window was broken.

In the month prior to my visit, local interest in the house had escalated. A large parcel appeared on its front porch and remained there for weeks. Then it disappeared. A local resident claimed she had seen smoke coming from the chimney. A few days later, another resident claimed the blinds in the windows had changed height. One evening, an elderly couple ran out of gas near the house. While waiting for help, they heard strange noises. Three nights later, a light was observed in a window. In the weeks that followed, several people observed "ghostlike figures" near the house at twilight. That settled matters for many of the local residents. Those who had been skeptical that the house was haunted were now convinced.

I checked the story with my grandmother. What I had heard and seen was accurate: "People in the town are scared, and so am I," she said, along while strongly cautioning me to "Stay away from that house!"

Several days later, the police searched the house and found no evidence of its occupation. There were no recent footprints or tire tracks on the grounds. No ashes were found in the fireplace. The local fuel distributor verified that he hadn't made a delivery for years. The electric meter was unchanged from previous months.

For a week after the police released their findings, there were no reports of unusual sightings. Then one evening, a highly respected local businessman and his wife drove by the house and saw "at least four ghostlike figures moving in and out of the bushes."

Three days later, the house burned to the ground. Evidence of arson was found. A less-than-thorough police investigation failed to identify suspects. The ghostlike figures disappeared for two years, and then they returned.

Attribution is one of the brain's systems for endowing attributes and explaining the behavior of oneself, others, and events.[26] It can be added to the list of mostly innate brain capacities, such as breathing and brain reading, which operate without perceived intent. At times it has been referred to as "agenticity."[27] It is one of the brain's ways for making sense out of the social, animal, and physical worlds. Its evolution among modern humans is assumed to be similar to that for imagining, belief, and brain reading: requisite systems were present among our ancestors and refined over time.

Attribution differs from brain reading and mirroring. Brain reading involves developing representations of the brain states of oneself and others that sometimes lead to beliefs. Mirroring is about the brain neurons of observers that are activated while observing others, which also sometimes lead to beliefs. Attributions are about assigning properties or causal inferences about why and how actions or events occur. Often beliefs and *divides* are in place when this happens. For example, "He quit his job because he hated his boss." "He converted to Catholicism to marry Jane." "The roof collapsed because of the weight of the snow."

Attributions may be internal or external. Those that are internal are assigned to one's personality and preferences and, more recently, to one's genes and possible inherited traits. People create clusters of attributes about themselves, believe them, and take them seriously. Such activity is associated with the belief that there is a self. For example, self-attribution is a likely explanation for the belief that God is personally involved in guiding one's life.[28] Groups often engage in similar self-attributions. Not surprisingly, how we see ourselves and others rarely map to how others see themselves or us.[29]

External attributions are those that are assigned to others, events, and contexts.[30] In their simplest form, external attributions work this way. From a variety of sources including direct and indirect evidence, brain reading, mirroring, beliefs, memory, and imaginings, person A develops representations about person B. The representations are then attributed—"projected," in a descriptive sense—to person B. That is, they become "properties" of person B. *Person A then interacts with person B or explains person B's behavior based on his attribution.* For example, person A attributes loving motives to person B and explains B's kindness to others as due to B's loving motives. Seeing what one believes is a form of attribution.

There is an obvious circularity to these happenings.

Attributions are not limited to humans. Typically those dealing with inanimate objects, such as automobiles, dishwashers, and lawn mowers, are more precise compared to those about humans. The workings of machines can be described in detail and causal explanations can be tested: for example,

"The lawn mower won't run because the carburetor is clogged." They also apply to animals. Folk-medicine views regarding the medicinal properties of animal parts provide a disquieting example of the consequences for species survival due to questionable attributions. For nonhuman primates, the number of species that are routinely sacrificed to obtain body parts are as follows: Neotropics = 19 of 139 species; Africa = 25 of 79 species; Madagascar = 10 of 93 species; Asia = 47 of 79 species.[31]

A reasonable guess is that among our ancestors, attributions worked very much as they do today. But differences are also likely. For example, two hundred thousand years ago, the range and variety of representations of others' behaviors and natural events were probably limited, compared to forty-five thousand years ago or today.

COMPUTATIONS AGAIN

Earlier in the chapter, it was suggested that the refinement of already-requisite systems didn't necessarily require increased computational capacities or energy expenditure by the brain for successful group living and mastering novel demographic challenges. Challenges include the migration of our ancestors into new environments, which led to the development of new technologies, increased population density, competition among groups, and increased within-group social complexity.[32] The refinement of brain reading and attributing capacities is compatible with this possibility. Mirroring may be an exception at least in crowded environments, such as a cocktail party or a political rally, where there is an increased chance of observing a variety of others and their behaviors leading perhaps to "mirroring overload" among observers.

The possibility of no increase in computational requirements may seem unlikely. But a possible explanation may be found in the gradual refinement and accumulation of beliefs and models that accurately predict outcomes. Crafting stone tools provides an example. Compared to the first attempts, later attempts would benefit from previous mistakes and learned efficiencies. In turn, craftspeople would more often select appropriate stones and apply efficient stone-chipping techniques.

An analogy can be made with athletic performance over the last fifty years in the sports of track and field. Literally every record has been broken, often significantly. It is improbable that these accomplishments are due to genetic change over a period of fewer than three generations. Rather, if the changes haven't been due to performance-enhancing drugs, they are probably a consequence of improved training techniques, training intensity, increased competition, expanded rewards for success, and increased numbers of participants. In effect, the requisite capacities were present fifty years ago. The demographics and challenges of the sports changed. The breaking of records followed.

A second analogy is to today's computers. A small cohort of talented individuals—every generation has some—was responsible for their invention and development. Once developed, their uses have expanded dramatically, even among individuals who understand very little about their inner workings. Their myriad applications have led to striking changes in the ways people socialize, conduct business, access and transmit information, and entertain themselves. All this has happened in only two generations without any likely change in the overall intelligence or computational capacities of *Homo sapiens*.

Of course it is possible to ask: Does it really matter if increased creativity and social complexity were contingent on an increase in intelligence or the refinement of requisite capacities? My answer is yes, it does. The refinement scenario is easier to reconcile with the amazing success of migrating groups to very different parts of the world.

14
STORIES AND MODELS

My research on vervet monkeys began in Saint Kitts. Subsequently, I traveled to West Africa, the likely spot from which centuries earlier the first vervets departed to the Americas on ships carrying slaves. Differences between vervets living in Saint Kitts and their West African counterparts might reveal changes that have taken place over multiple generations of separation. I hired two guides for the search. A canoe was rented and the three of us started upstream. Days one and two of the trip were unproductive. No vervets were sighted. On day three, we arrived at a small village.

The villagers—some of the children had never seen a "white man"—were friendly but cautious. We disembarked from our canoe and were escorted to an open space in the center of group of huts. The chief of the village—I'll call him Chief Fred—greeted us. Following several rituals, including drinking a hideous-tasting soup apparently to test if I could be trusted, the chief, I, and one of the guides who was an accomplished linguist talked. (The chief's statements below have been edited for brevity and clarity.)

During the conversation I asked him if a missionary lived in the village.

"Years ago, a missionary came to live with us. He advised us about how to behave and how to respect and worship his God. He promised that if we behaved the way his God wanted, we would be rewarded in this life and after. Our children would flourish, our food supply, which was often uncertain, would improve and stabilize, and we would find an inner peace. We believed his teachings and became Christians. We were baptized."

"Have events unfolded as the missionary said?"

"No. We soon encountered several years of famine. In another year, one-third of the village died because of disease. We sent the missionary out of the village."

"What's happened since?"

"We have devised our own rules. Children, mothers, and elders are fed first. Parents are responsible for their children. When we need to build a new hut, everyone participates. Those who don't cooperate are sent away. We are now healthier, we eat better, and there has been very little disease."

"What happened to the missionary's God?"

"Some of the elders still believe in him—the missionary was here fifteen years."

"Do you have another god?"

"No. Or maybe yes. Our god is our ability to survive and remain healthy. This we have solved for ourselves. The missionary's God is not for us."

Stories are accounts and explanations of personal experiences, events, imaginings, and hopes, like those of Chief Fred and the members of his village. Models identify patterns and explain events. We create stories and models and live in the world they record and explain.

Early forms of stories date back perhaps two hundred thousand years. This date is selected because stories usually require language, and language has an estimated history of two hundred thousand years.[1] Models, however, don't necessarily require language: a person lacking language could show others how to locate tubers, make stone tools, and catch animals. Models, thus, are likely to have a longer history than stories.

Whatever the historical details, the main focus of this chapter is not on the earliest moments of stories and models. About those moments, there are very few data. Rather, their possible histories are picked up approximately forty-five thousand years ago. At that time, language was established, and it's highly likely that both simple and sophisticated models were in use. Modern humans had migrated over much of the globe and had addressed multiple novel challenges. As they did, a reasonable guess is that their brains moved continually toward narrative and explanation—that is, they created stories and models. These would increasingly influence what they experienced, believed, and how they acted.

Migration to different areas meant that stories and models differed from one location to another. Many would have their origins in the unique demographic challenges that groups encountered. Hunting techniques, food preparation, shelter type, water crossing, social organization, and creation myths, all of which served as in-group markers,[2] were their likely topics. Together they would contribute to cultural myths, update the tool kits used for daily living, and stimulate cognitive and emotional explorations of imaginary worlds. By forty-five thousand years ago, it's likely that the supernatural had its own library of stories and was modeled in multiple ways.

Both stories and models exist as representations in the physical brain. That said, nothing about either of them suggests that they are products of specific evolved brain systems. Rather, they are products of multiple systems, such as memory, imaginings, brain reading, mirroring, attribution, and intuition. Moreover, the contributing systems are rarely the same for any two stories. This explains in part their variety and hints at their multiple social uses.[3] Models also are products of multiple contributing systems. However they differ from stories in that there are mostly innate and mostly learned models.

Most likely, stories and models emerged as bread-and-butter guide-lines for daily life much as is the case today. Likely also, they both narrowed and widened *divides* and gave structure to often chaotic experiences and disconnected bits of information, much as they do today. There is nothing surprising here: the brain has a strong tendency to organize and store information as beliefs, stories, and models.

STORIES

As with beliefs, stories serve multiple uses. They give structure and meaning to lives and events. Tellers and listeners are provided places in their social, physical, and metaphysical worlds. We relate to others through stories. They are platforms for relationships and reveal the intentions of their authors and their audiences.[4] They can serve as a form of knowledge, gossip, and rumor as well as a medium for persuasion.[5] They are the history books for illiterate people and the basis for much folk psychology. Their content is limitless and they often become beliefs.

The reasons for telling stories vary as much as their uses. Personal stories deal with the lives of those who tell them, family, how others believe and behave, and weird happenings. They often entertain listeners and focus attention on the storyteller. When they do, they are satisfying to those who tell them.[6] At times, they are told to change the direction of a confusing or an irritating conversation.

Yet, as important as they are, personal stories have no life of their own unless they are told again and again.[7] They are highly vulnerable to extinction. As an example, take the fate of the stories you heard from your grandparents when you were young. As an adult, some may be recalled and told to your children. Very likely, fewer will be heard by your grandchildren. Like languages that become extinct and thereby erode knowledge, so too for stories that are no longer told.[8] While this may seem an obvious point, it has troubling implications: when stories become extinct, part of the brain's library of its owner's personal history and culture is lost.

There is a further factor that contributes to extinction. Personal stories are often filled with a teller's emotions. Memories of emotions tend to fade and lose their valence while cognitive details tend to have a longer life. Often a listener can recall many of the details of another's story but less of the teller's or his own emotions when a story was told. There are, of course, partial remedies to this situation. Before printing, there were group storytellers, essentially keepers of important group-relevant stories who often infused emotion into their tales. With printing and historical scholarship, the preservation of past events improved, yet emotions often failed to find a place on the printed page. Moreover, even with printing, preservation of the past is far from complete. For example, how many histories of the United States discuss the Bavarian Illuminati Conspiracy that ignited intense feeling among the population of New England in 1798?[9] Or how many note that during the nineteenth century in both England and the United States, there was a thriving profession of spiritual photography depicting images of dead loved ones who had returned from other worlds?[10]

Not unexpectedly, many factors influence how stories are structured and told. Details may be condensed or elaborated. Events may be skewed with the intention of creating a favorable impression among listeners. At times what is told is knowingly crafted to hide a teller's motive, which has a distinct fMRI neural signature.[11] Stories may be inaccurate because of a teller's self-deception. Or inaccuracy may be due to the effects of aging.[12] There are also unintended editing effects when stories travel from brain to brain: there is a parlor game in which people sit in a circle and, as accurately as they can, whisper one to another a story that begins and ends at the head of the circle. After the story has traveled around the circle, it is often unrecognizable from its original version. There are variations, of course. For example, religious stories and myths are partially counterfactual and partially counterintuitive, which, as noted earlier, may make them easier to remember.[13]

Telling a personal story involves communicating one's experiences in ways that others can understand. This requires translating one form of information to another form. This often works successfully. For example, we may flinch on hearing of a terrifying incident, but the point to note is that we are flinching at what is communicated, not at the incident itself. The communication has activated our emotions.

Do stories of one's experiences accurately depict events? Probably not. Why? Because personal experiences rarely occur in organized ways. Stories of these experiences involve post hoc reconstructions that give them structure and improve their likelihood of being understood. Said another way, the process of turning an experience into narrative assures that the narrative will differ from experience no matter how serious a storyteller's efforts to communicate accurately.[14] Writers, perhaps because of the relatively slow pace of writing compared to speaking, are particularly sensitive to this point. Emerson, Melville, Eliot, Joyce, Proust, Lawrence, and Borges all struggled to overcome translation obstacles but never did so to their full satisfaction.[15] Assessing how an audience might respond to a story is also a factor: certain stories are either highly modified or not told to certain ethnic, religious, professional, or young audiences.

The other half of storytelling is that a listener must interpret what he has heard. Understanding a storyteller's intended interpretation is contingent on both the teller and a listener sharing similar views. Telling a friend of my frustration while waiting forty-five minutes to deposit a check at a local bank will be understood provided my friend can put himself in my shoes (Theory of brain) and imagine his own frustration in a similar situation. A listener's "I don't get it" signals the absence of a shared view.

Jokes—they too are stories—provide an informative example of shared views between teller and listener. Certain jokes are effective when the storyline entices the listener to expect one ending but the story provides another. For example:

A man is driving down a country road and sees a farmer walking with a three-legged pig. He stops his car, approaches the farmer, and asks: "Is that your pig?"

"Yes," replies the farmer.

"He has only three legs," comments the man.

"Right," replies the farmer.

"How did he lose his leg?" asks the man.

"Let me tell you about this pig," the farmer replies. "Last year, when my wife and I were asleep, a lamp tipped over in the house and started a fire. The pig detected the fire, broke through the back door, raced to our bedroom, and woke us up. We're alive today because of that pig. And just last month I was alone on the farm when the tractor tipped over and pinned me to the ground. The circulation to my legs was cut off. I thought I was a goner. The pig assessed the situation, raced to town, alerted the police, and they came and rescued me. I have my legs and can walk today because of that pig."

"That's amazing. What a pig!" replies the man. "But you haven't told me how the pig lost its leg."

To which the farmer replies: "You wouldn't want to eat a good pig like that all at once."

The joke works because the structure and content of the story suggest another ending, such as that the pig lost its leg in an accident or was born with one leg missing. Nothing in the story hints that the farmer might eat the pig and especially so, given its exceptional qualities or, worse yet, eat only part of the pig from time to time. In effect, the anticipated ending on the part of the listener is mistaken. It's the recognition of the mistake that is the basis for laughter.[16]

Multiple factors influence how stories are created. Soon-to-be-discussed models deal with how the world works, and they serve to organize and structure experience as well as discard irrelevant bits of information. Stories often reflect their organization and structure. Mirroring may serve to initiate a story in response to another's state: for example, observing a person in pain may remind the observer of a pain-related story. Attributions often provide story content. And stories and models can contribute to attributions and brain reading. Most often they reduce *divides*, although they can widen or render them indeterminate. On the other hand, intentional *divide* widening is a favorite pastime of political activists and competitors who are intent on discrediting others.

Stories are associated with action. People who believe their own stories often view them as causing their behavior. The stories of others may work the same way. Those about gold and fortunes to be had in places like Timbuktu have been associated with action among those who heard and believed them.[17] In effect, stories have their own authority, and it's the brains of both tellers and listeners that create and grant the authority.

How might the preceding points apply to forty-five thousand years ago? The likely answers are that stories increasingly contributed to the symbolic and technological complexity of the period, furthered social bonding, and intensified both in-group and out-group identification. They would codify much of the expanding range of human experience that accompanied migration, technological and cultural change, and increasingly sophisticated socialization. This suggests that there should be a positive correlation between the age of a culture and the number of stories known to its members—a kind of continually enlarging encyclopedia of the events and experiences of daily life, even though simultaneously some stories were becoming extinct. Many of these stories would contain messages about what to do or not to do or about heroes. Surely our ancestors had the equivalents of current stories such as the little boy who cried wolf, Paul Bunyan, and werewolves. Surely also there were stories about gods and the workings of mysterious forces of the universe.

MODELS

In the most general sense, models are anything used to represent something else. They may be likened to explanations, formulas, archetypes, or templates about how the experienced and supernatural worlds work. Unlike many stories, models often are free of emotion.

There is an extensive catalogue of model types: metaphysical, mathematical, logical, economic, data, and so forth. None of these are topics here. What follows is a discussion of models that we use in daily life—that is, models that people believe or that they use even if they are not believed because they are all that is available. They are key ingredients in the tool kits for daily living, ranging from how to fry an egg to managing a business efficiently. Rarely is a single model adequate to fully explain a phenomenon. Rather, several models are often present dealing with the same evidence or contingencies—for example, models of social behavior are often used to explain economic behavior and vice versa.[18] There are models that are mostly innate and those that are mostly learned.

It can't be emphasized enough how critical models are for managing the activities of daily life. There are literally hundreds of such models associated with almost everything we do: not changing the oil in your automobile will lead to motor damage, thoroughly chewing food prevents indigestion, good examination grades in high school will get one into a good college, and careful techniques produce good surgical outcomes. We believe them. As they become more precise and predictive of outcome, a reduction in computational requirements is likely coupled with their conversion to beliefs.

MOSTLY INNATE MODELS

Multiple studies of child development have established that children can think before they speak,[19] and they are born with the capacity to use models to deal with how the world works. For example, by sixteen months of age, they can infer the causes of failed actions:[20] if a light switch is flipped and the light doesn't go on, it could be due to selecting the wrong switch or a broken bulb. Children can figure out the most likely option. They also have concepts of property rights[21] and can grasp abstract geometric principles despite the absence of math training.[22] They apply cause-and-effect models in their interpretations of events that are closely related in time, such as noise from the kitchen at dusk is due to preparation for dinner. In short, children enter the world with strong predispositions to process information in specific ways.[23]

Model-related influences on behavior may extend throughout life. A person experiencing the sudden onset of anxiety while speaking for the first time to a large audience may avoid speaking in such a forum again. Exposure to high places accompanied by fear that one will fall or might jump often leads to the subsequent avoidance of heights. The presence of pain following physical activity may lead to a change in work or exercise habits. Models appear to be critical for intuition, especially when they are the source of near-instant explanations of complex events or information.

Mostly innate models are so fundamental to the way we think that they are not easily unlearned. This is due in part to their predictive accuracy and utility and in part to the absence of compelling alternative models for much of what is experienced in everyday living. To maintain that coincidence explains two events that are closely connected in time is seldom convincing. Still, within limits, mostly innate models can be modified. Adding a temporal feature separating cause and effect is an example. The child who plants his first radish seed may expect signs of growth within hours. With experience, he will expect seeds to sprout after a week or two.

There is a long list of familiar variations of cause-and-effect models. A sampling includes the following:

(a) *Repeat behavior associated with pleasure.* Children act in ways that are consistent with this model. They need not be taught that ice cream is a source of pleasure: a model makes the connection. Information may have a similar outcome. The sight of food, for example, may be followed by a desire to eat. As mentioned, experiments show that brain activity in areas associated with pleasure and reward precede actions that have a history of pleasurable responses.[24] This may explain why disregarding the behavior-influencing effects of mostly innate models associated with pleasure is so difficult. (One downside of pleasure-based associations is that they appear to be a significant contributing factor in drug addiction, high-risk behavior, and behavior perversions.)

(b) *Avoid behavior that is associated with displeasure or the absence of reward.* Children act in ways that are consistent with this variation.

(c) *Repay those who are the cause of personal injury.* While emotions and plans for retaliation are not uniformly followed by action, they rarely disappear overnight. Formerly married individuals who consider themselves victims of infidelities of their ex-spouses frequently fit this model. So too with those who have experienced insults, pain, or betrayal at the hands of others. A personal experience catches the flavor of these situations.

During my senior year in college, I lived on the third story of an apartment building. There was an empty apartment next to mine. A new tenant was expected soon. One Sunday morning, as I departed from the apartment to get a donut and a cup of coffee, a car with a large hooded trailer arrived. I suspected it was the new tenant. I introduced myself. He did likewise. I asked if I could help him move in. He accepted. Two hours later and after our thirty trips up three flights of stairs, his belongings were in his apartment.

Several months passed, during which we encountered each other infrequently. Then the time arrived for me to vacate my apartment. In anticipation of the move, I knocked on my neighbor's door. When he opened the door I asked if he "would help me move the following Saturday?" "I don't return favors" was his reply as he shut the door. I was consumed with anger.

It is now decades later. Each time I recall the incident, I still fill anger. And to this day, I wish I had responded differently.

(d) *An intimate relationship with an attractive person will be pleasurable.* While positive and inviting feelings about interacting with attractive persons are most intense during the teens and early twenties, they continue throughout life and are difficult to modify.

(e) *Members of groups other than one's own are dangerous.* If this model isn't mostly innate, how can the near-universal distrust and competition among groups be explained?

None of this suggests that mostly innate models accurately depict how the world works. Nonetheless, they tend to narrow *divides* and all the more so when they consistently predict outcomes. The more they do so, the more likely they become beliefs.

MOSTLY LEARNED MODELS

It has been known for nearly a century that, during development, children change their model of balance scales—teeter-totters—several times. From an initial focus on weight to a later consideration of the distance dimension, they eventually settle for a model incorporating multiplication of the relevant factors affecting balance.[25] As simple as a balance scale may seem to adults, it wasn't always so. Models that develop this way are sometimes referred to as "bottom-up" models. These are applicable to learning selected skills and involve taking procedural or practical knowledge and transforming it into conceptual knowledge—that is, tool-kit-related knowledge. Such models contrast with "top-down" models in which conceptual knowledge is translated into procedural knowledge. This happens, for example, when receiving instruction about how to use a complex machine. Either way, the outcome may be similar: *once models are learned, they are used by the brain to interpret information.*[26] And, typically, the steps of learning are usually lost to recall: for example, try recalling your first focus on weight as an explanation for the behavior of teeter-totters.

No doubt many learned that bottom-up and top-down models merge. This happens, for example, in learning to play tennis: if the trajectory of the tennis ball is other than one predicts, adjustments in one's footing and tennis stroke follow, but within the options and constraints of the game's rules. More technically, working memory is first transferred to movement and model development.[27] This is followed by modification of models in response to model-variant experiences, a point that is predicted from studies that show characteristic neural activity in response to new information.[28] For both bottom-up and top-down models, practice is critical for refinement. It offers the advantage of direct evidence regarding model predictability and may lead to model revision and use.

Another type of learned model is analogous to an algebra equation: the rules for its application are specified, but what the symbols represent may differ. For example, the model $A = B + C - D$ can apply to who is and isn't invited to a dinner party, a new business plan, a recipe, or a design for a military action in a distant country. At times, revision occurs following model testing, such as tasting a meal cooked according to a recipe designed to achieve a special flavor and texture.

Then there are idiosyncratic models, such as those developed by families dealing with how they should behave following the death of kin: $A = B + C + D + E + F \div G - H$, in which the symbols refer to such things as kin coming together, family members showing remorse, a funeral, eulogies, and burial. Most but not necessarily all of the variables in the model may be honored. Idiosyncratic models based on imaginings and beliefs are also active in daily life. The world of spirits, gods, angels, and animals with humanlike motives and personalities provide much of their content.

Not surprisingly, the manner in which mostly learned models organize and structure behavior is subject to multiple influences. This occurs because imaginings, beliefs, brain reading, possibly mirroring, and often considerable trial and error differentially contribute to model development and modification. Nonrational features of reasoning and belief, such as those discussed in chapter 4, frequently complicate matters—apparently there is no standard formula for development and modification. Nonetheless, many such models possess a common property: they are attributed to others, animals, inanimate entities, or the supernatural in the process of explaining behavior.

In what ways do models fit in the scenario of happenings forty-five thousand years ago? The likely answer is similar to that for stories: they increasingly contributed to symbolic and technological complexity of the period as well as beliefs. For mostly innate models, there may have been few differences from those of today. For mostly learned models, it's highly probable that their number, complexity, and degree of refinement have increased significantly—their utility has been recognized. Moreover, at the present time, they are likely to outnumber mostly innate models: new mostly innate models presuppose genetic change, which is slow, while new mostly learned models can be created in moments.

15
TRIGGERING

Before my first trip to West Africa, I visited the Los Angeles Zoo to familiarize myself with animals I might encounter. Both black and green mambas were of special interest. Lore had it that no one had ever survived a bite from one.

After spending ten minutes looking through the glass-enclosed cage labeled "Black Mamba," I couldn't find a snake. Eventually a zoo employee came by and I asked for assistance. She pointed to the snake. It was resting on a branch of a tree with its head extended some eight inches beyond the end of the branch, thus appearing as an extension of the branch. If I couldn't identify the snake two feet away in the cage, I wondered what my chances would be in the wild.

For the following two weeks, I had a recurrent dream. I was somewhere in Africa, preparing to go to bed. I would take off my shirt and hang it on what appeared to be a branch of a tree. The branch, of course, was a black mamba, just as in the zoo in Los Angeles. Each time I tried to hang the shirt, it dropped to the ground below the branch. I would pick it up and hang it on the same branch again. Through all this, the mamba was becoming increasingly irritated and prepared to strike. Just as it was about to strike, I would wake up anxious and sweating. After two weeks, I no longer had the dream.

A month later, I was in Africa. Throughout the trip, whenever I saw a branch that resembled what I had seen in Los Angeles, I experienced a moment of anxiety followed by very careful vigilance. The branch served as a *trigger* for a specific response both emotionally and cognitively. (I never saw a black or a green mamba on the trip.)

Triggering occurs when either external or internal information (stimuli) initiates brain activity and specific states of awareness. It is a special case of the *extended mind hypothesis*. It is so commonplace and frequent that we rarely notice that it is occurring. It has an interesting if not profound implication however: it's not quite that we are walking around the world as "freethinkers" and "impassionate observers"; rather, to a significant degree, what we experience in awareness is determined by the environment in which we find ourselves.

INNATE RELEASING MECHANISMS

The concept of triggering was developed by Nicholas Tinbergen during his studies of animal behavior: a specific stimulus—information—elicits a predictable behavioral response in another animal via its effects on the animal's *innate releasing mechanisms*. Such behavior is definable by its form and structure. It is species universal in that innate releasing mechanisms have evolved to respond to specific releasing stimuli.[1] Releasing stimuli can be cumulative when more than one stimulus is required to initiate a response, or they may be additive, as can occur when the same stimulus is repeated.[2] Typical behavioral responses among animals include escape, prey catching, courtship, and fighting.

At times, animals behave in ways that suggest that their behavior is initiated by innate releasing mechanisms. This happens when dogs are trained to follow commands such as "sit" and "fetch." Certainly dogs don't know the meaning of these commands when they are born, nor are they versed in the many languages and thus different stimuli in which such commands are given. Nonetheless, with training, commands are followed. Brain mechanisms associated with learning account for this type of behavior, which is not species universal.

HUMAN PARALLELS AND SIMILARITIES

Although some scholars question the presence of innate releasing mechanisms in humans, a reasonable case can be made that they exist.[3] Examples include reflexes, such as blinking one's eyes in response to the flash of a bright light and withdrawal from a painful stimulus. Mirroring also may qualify: a specific behavior by others initiates predictable neuronal activity among observers. Recently it has been postulated that information present in the brain, such as an image of God, can serve as a stimulus that initiates the nonvocal aspect of petitioning prayer.[4]

Less straightforward examples are found in stimuli such as loud sounds, disgusting smells, and inviting sexual postures that initiate specific responses. Such responses occur frequently, but they can be modified and don't appear to be species universal. (The fact that advertisements often include pictures of attractive women and men or locations that are unrelated to what is advertised suggests that advertisers believe that human purchasing behavior can be triggered—recall the discussion in chapter 5 dealing with neuromarketing.)

Then there are behavior responses among humans that are primarily learned yet may be highly predictable. For example, people indoctrinated with religious stimuli early in life frequently retain their responses throughout their lives. For these individuals, a picture of a religious figure or listening to a hymn triggers a specific behavior and emotion. But religious indoctrination is not unique. A similar point applies to well-known friends, disliked enemies, infants and parents, and spouses: for example, the posture, the voice tone, and what is said by spouses often trigger specific states of awareness. Much the same applies to bosses and employees. Such responses may be viewed positively in that they facilitate social communication, job efficiency, and sensitivity to others. But there are other considerations.

RESPONSES TO TRIGGERING

Among humans, responses to social stimuli might develop under the following conditions. (1) A high percentage of the social stimuli of daily life are similar and predictable. (2) Predictable responses to these stimuli facilitate social interaction. (3) Predictable responses reduce the brain's computational requirements and energy expenditure.

That responses to triggers under condition (1) are primarily learned seems clear, although a well-functioning memory is essential. Studies show that repetition improves memory—recall learning the multiplication tables—and increases the likelihood that a specific stimulus will lead to the retrieval of a specific response.[5] How this might take place in the brain is suggested by studies showing that memory repetition is associated with neural pattern similarity, which enhances episodic recall to the same stimulus.[6] And, as noted earlier, the resting brain recapitulates activity patterns that occur during recent experiences, which appear to contribute to memory longevity, specificity, and preparation for similar events in the future.[7]

Condition (2)—predictable responses to social triggers facilitate social interaction—also appears to be primarily learned. Such responses should considerably reduce the potential for ambiguity that can accompany such interactions.

Condition (3)—predictable responses reduce the brain's computational requirements and energy expenditure—is consistent with the view that brain systems are biased in favor of conserving energy.

Although it is seldom viewed this way, triggers are a type of brain control. If I respond in a predictable way to a verbal stimulus, the person responsible for the stimulus briefly controls activities in my brain. In familiar situations, undesirable consequences rarely follow. However, as is discussed in upcoming chapters, the effects and consequences of this type of brain control can be significant.

The preceding points add up this way. When persons live in predictable social and physical environments, a large percentage of the stimuli they encounter are similar in form and structure. They respond to these stimuli in highly predictable ways. As a result, computational requirements dealing with such interactions are reduced. A rough estimate is that 40 percent of the social interactions that occur in daily life meet these criteria—this percentage hints at why it is stressful being in an unfamiliar social setting or physical environment. Two points follow. The prevalence of triggering and predictable responses is a rough measure of the degree to which humans are socialized and influence each other. The effects of triggering are exceedingly difficult to avoid or control.

There are other types of triggering.

The two guides who took me up the river where I met Chief Fred had very different talents. One was a natural linguist and spoke at least half a dozen local dialects along with English and French. He was the translator in the meeting with the chief. The other was an expert in the environment and its animals. For most of his thirty-five years, he had lived in the bush. He excelled as a hunter and in finding the occasional person lost in the forest. As we trekked through the forest in search of monkeys, he was our guide. His capacities seemed uncanny. We would walk for a time and not see an animal. The usual sounds were those of birds, the wind, and falling branches. Then, unexpectedly, he would hold up his hand, point to an area, and whisper, "Monkeys are there." Invariably they were. Neither the linguist nor I were able to match his skills.

After several years in Saint Kitts, I began to notice that I could do much the same as the guide. Novice guests and funding-agency administrators who had spent most of their lives in ten-by-fifteen-foot offices and visited the island to see the monkeys were astounded at my "special skill."

But is it a special skill? I doubt it. Rather, the skill is learned following thousands of experiences in which combinations of factors are associated with the presence of monkeys. The wind is one factor: the less the wind, the more likely monkeys. The density of the foliage is another: the greater the density, the more likely monkeys. The activity of non-primate animals in the area is yet another factor: the greater their activity, the more likely monkeys. After a while, it is unnecessary to systematically wander through the list of factors. Rather, combinations of factors lead to probabilities or estimates of the presence of monkeys.

EXAMPLES

Triggering is far from an abstract idea. It is a very real phenomenon for all of us. And triggers are types of energy. A few words or nonverbal cues are often sufficient to initiate predictable behaviors and states of awareness. For example, when a friend asks you "How are your children?" information about your children pops into awareness without obvious effort on your part. The question triggers the brain systems responsible for what you experience. Systems and information for responding to familiar stimuli are in place. The response affirms that information about your children is already present and organized in representations. You know your children. You believe what is present in awareness. *Divides* are unlikely to be present. You might believe that you thought of the answer to the question, but your belief occurs after your response is initiated.

There are other kinds of responses. Say a person to whom you have attributed devious motives asks, "How are your children?" Rather than awareness of your children, you are likely to experience concern about why the person is asking the question. What you experience is influenced by what you have attributed. You may not respond.

These examples differ from those in which there is no ready response to a trigger. Suppose a friend asks, "Do you think your daughter would enjoy a year abroad in India?" An immediate response is unlikely to appear in awareness. There is no existing representation. What you are likely to experience is some of your daughter's attributes and what you know about India, following which an answer appears. In effect, you generate a new model, perhaps also a belief and a *divide*, which could range from narrow to wide. The presence of the model would permit a rapid response if the question is asked at a later time. But again, the systems responsible for both the development of the model and your response are outside awareness.

Similar scenarios apply to responses triggered by animals. For example, even at a distance of ten yards, most people respond with fright when they encounter a coiled rattlesnake shaking its rattle. The snake triggers their response. Such responses result from attributions that rattlesnakes are dangerous and they bite, which of course is correct. However, in the same circumstance, seasoned hikers will be relaxed because they attribute differently: rattlesnakes do not strike unless their spatial territory (which is about one yard from their head, if they are coiled) is violated. At a distance of ten yards, there is minimal danger.

Triggered responses can occur in response to literally any information. If asked what you think about the president, you are likely to experience a ready answer. Music works the same way. Different types of music lead to predictable emotional responses among listeners and, as might be expected, fMRI studies reveal that music activates nearly every part of the brain.[8] Familiar choices work the same way. Recall the examples in chapter 10 dealing with the selection of an entrée while dining: the items on the menu serve as triggers, which initiated predictable brain activity. Or what happens when someone announces that he will tell a lackluster joke that you have heard several times before?

At other times, triggers and responses are highly ritualized. Consider a commonly observed interaction between two friends encountering each other while walking in opposite directions.

John: "Hi, Bill. How are you?
Bill: "Fine, and you?"
John: "Great; see you soon."

The John-Bill interaction might appear to convey minimal information. But, in fact, it contains important information for both John and Bill. Bill's response of "Fine, and you?" signifies to John that their friendship continues and no unexpected surprises are pending. His tone of voice may confirm his message.[9] The brevity of such interactions is consistent with the idea that those who are interacting have conveyed and obtained the information they seek, although this is not obvious in the verbal content of what is said—in effect, a familiar response to a trigger is worth a thousand words. This brevity contrasts to situations in which what is said is interpreted literally as in, for example, encountering an acquaintance walking in the opposite direction who spends fifteen minutes responding to, "Hi. How are you?"

At times, gestures take the place of words. Among friends, an elevated thumb, a special smile, a head nod, or a raised eyebrow can each serve as a response to questions such as, "How are you?" These types of interactions are brief and contained—a kind of communication shorthand. They are not thought through.

There are multiple influencing factors. Responses may be affected by mirroring if one also is responding to the facial expression or posture of the person who is asking the question. The age of a person may trigger a specific response. Or the trigger may be a military or police uniform. There are also cultural factors: on first visiting an unfamiliar culture one lacks already-available responses to even routine greetings or questions, and *divides* dealing with what might be appropriate responses are usually wide.

Essentially the same points apply to internal stimuli. A pain in the leg is likely to initiate a model of a cramp, perhaps also its remedy. Familiar mechanical-related stimuli are handled much the same way: a flat tire, running out of gas, a lawnmower failing to start, or the lights going out trigger explanations of their possible causes as well as tool-kit solutions.

Triggering even applies when only indirect evidence is available. News that North Korean negotiators have again walked away from the negotiation table, that members of Mexican drug cartels have been involved in gun battles, or that Israel and the Palestinian Authority are unable to agree on an issue are followed by familiar beliefs and models in awareness.

A common feature of the examples above is that the response to a trigger can quickly meld into a belief. You are likely to believe that the friend who asks about your children is interested in your children. For the non-friend who asks the same question, you are likely to believe that he has an unstated motive for doing so. When you relax a painful leg and the pain disappears, you are likely to be convinced that the source of the pain was a cramp.

THEMES

Several themes interweave through the preceding discussion. One is that the brain has evolved biases favoring predictable, low-energy-expenditure social interactions. Triggers and predictable responses are consistent with this view.

A second theme is that external and internal stimuli have far more influence on our brain activity and awareness than we recognize. That this influence is ever present perhaps explains why efforts at serious concentration require removing oneself from triggering stimuli.

Third, the brain has distaste for unpleasant triggers. An illustration of this point comes from a conversation I had with the manager of a radio station in Los Angeles. The station advertised that it covered the news of the world, including local traffic and weather, every twenty-two minutes. Meeting this time schedule required that the announcers gloss over literally every item they reported, except for the time of day. The manager had adopted the format after he discovered that reporting on the details and ambiguities of many news items resulted in a noticeable decline in the number of listeners to the station. He queried listeners about reasons for the decline. A typical reply was that they listened to the station "to be assured that nothing unusual, frightening, or disastrous was happening." In Los Angeles, reports of traffic jams or smog levels are familiar and irritating but not frightening triggers. However, reporting on the probability of the next major earthquake or the possibility of another 9/11 had aversive emotional effects among listeners.

Fourth, triggering effects can spread through groups much like a plague. *Belief contagion* is a term often used to characterize such events. It describes how, in response to a trigger, members of groups can come to share a common belief and emotion with no discernible *divide*. Often there is a "priming" effect due to people already sharing similar mental states.[10] A trigger such as an incident of public abuse or a perceived wrong decision during a sporting event serves to catalyze a sense of like-mindedness among participants, perhaps also a shared view of a common destiny.[11] Action often follows. Spontaneous behavior seen at political conventions and antiestablishment demonstrations are examples.

Fifth, triggering has direct physiological and structural effects on the brain. This is not surprising. Previously mentioned examples include changes in amygdala activity in response to threats and threat-related postures, the effects of music on the brain, and stressful work conditions. Structural effects are noted in the positive correlation between the volume of the amygdala and one's social-network size: the more friends one has and with whom one interacts, the larger one's amygdala.[12]

TWO HUNDRED THOUSAND TO FORTY-FIVE THOUSAND YEARS AGO

If responses to triggering in social interactions are largely examples of learned responses, their history among humans likely dates back well before the emergence of culture and possibly to the common ancestor of today's primates. Certainly nonhuman primates exhibit such behavior—recall the trained monkeys that perform as commanded by the organ grinder. Certainly also mothers and infants developed familiar and predictable social triggers and responses. Learned responses are also likely in dealing with environmental information. As people became more familiar with the environment, predictable responses to events such as encountering a dangerous animal followed. And, as groups enlarged and became more cohesive, the more likely it was that a greater percentage of their social interactions involved familiar triggers and responses.

16

INTRANSIGENT BELIEFS AND BELIEF-DISCONFIRMATION FAILURE

"People act themselves into a way of believing as readily as they believe themselves into a way of acting."[1]

BACK TO SAINT KITTS

On my forth trip to the island, I encountered a young man—Thomas— who had just returned home from completing college in the United States. He came to where we lived seeking a job. Part of our conversation follows.

> *Author:* "You really think you would like studying monkeys?"
>
> *Thomas:* "Yes, I've seen them all my life, but I never paid attention to their doings."
>
> *Author:* "You realize that the current studies require that we get up at three a.m. so we can locate ourselves in the hills before the monkeys get up. And it gets hot out there. We stay and observe until three or four in the afternoon; then there is the walk back over the mountain."
>
> *Thomas:* "I'm young and I can do that. But there is another reason. I want to know what you do for the CIA."
>
> *Author:* "The CIA?"
>
> *Thomas:* "Yes, the CIA."
>
> *Author:* "That's a first. What makes you think I work for the CIA?"
>
> *Thomas:* "I think you all do, not just you."

Author: "OK, but let's get back to my question: Where did you get this idea?"

Thomas: "It's obvious to me. You come and go from here regularly. You have all the new equipment. At the telegraph office, they say you send messages frequently and many go to your government. And then there are the monkeys. You tell people they are of scientific interest, but they're just pests—everyone on the island knows that. They eat our crops and run away. Nobody would give you money to study pests. That's crazy, like studying rats or bedbugs. I've heard they carry deadly diseases. That's why most people won't eat them. Some people think that you want to catch them and study their diseases so the CIA can use them. There is another matter. Last week, a government plane landed here and a person came to see you."

Author: "Thomas, you're amazing and your ideas are amazing. Do others have the same idea?"

Thomas: "Yes, people know you work for the CIA. Studying the monkeys is just a front."

Author: "Is there anything else we are doing for the CIA?"

Thomas: "I believe you are mapping the island."

Author: "What?"

Thomas: "Yes, mapping. You now know more about the island than most anyone who lives here. You have been to the volcano many times, just like you've been to the peninsula, where no one ever goes. Maybe there will be an invasion and you will have all the information."

I checked with other locals I knew on the island. They agreed: we were a CIA front.

I hired Thomas, who turned out to be an excellent monkey observer and employee. Together we went over the daily data sheets, discussed how scientific papers would be written, wrote them, and sent them off to scientific journals. He became a member of our research family. He knew how we lived and what we thought. Our work and lives were transparent. I must have explained to him at least a dozen times that our telegrams to the government were to that part of government that funded scientific research and that the CIA was an entirely different part of government. And I explained to him again and again that the people who came to see us were either scientists or people from funding agencies who were checking on our research. And I explained and explained how the monkeys were of great scientific interest.

Eighteen years later, when I left the island for the last time, Thomas came to see me off. I asked him if he had changed his mind about the CIA. "No, but I have come to like the CIA because I like you" was his farewell reply.

Some beliefs are intransigent. They change rarely.

One type is those for which there is and has been near consensus for as long as anyone can remember. Water flows downhill, not uphill, is an example. These beliefs are associated with justifying evidence. They consistently predict outcomes and they can be tested.

Another type includes beliefs for which there is an absence of justifying evidence. They predict outcomes only to those who believe them. Examples include that there are gods or higher powers, people are basically good or basically bad, science is the sure antidote for all misunderstanding, political and idiosyncratic ideologies map to reality, and that I and others on Saint Kitts were employees of the CIA. This type is the topic of this chapter.

INTRANSIGENT BELIEFS

Resistance to change is the default state of beliefs in which people are emotionally and cognitively invested. Intransigent beliefs are preeminent examples. To those who adopt them, they convincingly explain what they believe. Ambiguity and uncertainty are reduced significantly, and pleasurable physiologic and psychological states are usually present. Once established, they require minimal energy expenditure by the brain. Their *divides* are narrow.

Those harboring such beliefs assert them with conviction and insist that they are buttressed by confirming evidence or authority. Alternative beliefs and evidence are disregarded or rejected usually via negative attributions. Those committed to alternative beliefs represent a dangerous opposition. *It is the refutation of alternative beliefs and evidence combined with the identification of those who harbor such beliefs or evidence as enemies that are signature features of intransigent beliefs.*

Belief perseverance is a term often used to describe intransigence. It refers to retaining a belief in circumstances in which there are equally or more compelling beliefs or clear evidence that is at odds with a belief to which a person or a group is committed.[2] Deeply held political ideologies qualify. Believers are convinced of their relevance and worth as well as the wrongness and irrelevance of alternative political views. Individuals committed to religious beliefs—atheism too—often believe much the same way. Those with other beliefs are either uninformed or enemies of the truth.

There is an interesting question for which there is no known answer: What is the percentage of people's beliefs that are intransigent? If those for which there is consensus and justifying evidence are excluded, a guess is 50 percent. On first past, 50 percent seems an unreasonably high percentage. Yet it seldom appears so when listening to discussions dealing with politics, human nature, religion, climate change, evolution, homosexuality, abortion, same-sex marriages, and moral issues. As often as not, such conversations amount to a series of belief-related assertions that undergo minimal if any revision. Recall that only two of the forty people interviewed for chapter 7 had changed a "deeply held belief" in the preceding six months.

How do such beliefs develop? Strong candidates include indoctrination by others, self-indoctrination, culture, experience, and imaginings that turn to beliefs.

INDOCTRINATION BY OTHERS

Indoctrination amounts to inculcating individuals with specific ideas, attitudes, emotions, models, values, or cognitive strategies—in short, beliefs and models. These give structure, purpose, understanding, and meaning to experience, ideologies, supposed evidence, and myths in highly specific ways. Indoctrination amounts to teaching others to respond in predictable ways to specific triggers—that is, it's a form of brain control. The process can be distinguished from liberal education, where questioning what is taught is acceptable and encouraged.

There is an absence of self-reflection. The ambiguities and uncertainties that trouble those who attempt to understand and explain human behavior or the physical world are absent except for an occasional verbal dressing.

There is no meaningful dialogue with those holding different views or evidence. No dialogue results from the absence of shared or at least similar beliefs, models, and evidence from which dialogue can proceed. The indoctrinations of children brought up in religion-committed families or schools are familiar examples. The details of what is taught differ across families and schools, but those who teach share the aim of molding belief, emotion, and behavior.[3] Ritual is often an important part of the process in that it reinforces what is believed. Christianity, Islam, Judaism, Hinduism, and their many variants have specific rules regarding behavior, thought, preferences, prayer, votive offerings, and the observance of sacred moments.

That indoctrination works as well as it does is not surprising. The brain is prepared to participate in the process. There are brain substrates of long-term-memory conformity that are responsive to social manipulation and have an identifiable fMRI brain signature.[4] They reveal a strong tendency of memory to conform to group recollections independent of the accuracy of what is recalled. When carried out successfully, those who are indoctrinated embrace the view that their beliefs are superior to those of others, particularly those of possible distracters. The net effect is that doctrines serve as blinders to alternative views.

It is now clear that indoctrination can take place among adults as well as children. People of all ages are susceptible. In part, this is the story of adults who are converted to religions, political ideologies, secular causes and agendas, and a portion of prison-camp captives. Groups commonly known as cults provide clear examples. The Branch Davidian sect, mystery cults of Asia such as the Many Faces of God cult,[5] the Aum Shinrikyo cult in Japan,[6] and Jonestown—to name only a few—are religious-political indoctrination systems. At times, their influence has been so pervasive that it overrides evolved tendencies to survive and reproduce. For example, 530 members of the Ugandan cult Restoration of the Ten Commandments died in a mass suicide in March of 2000,[7] and a similar carnage occurred in Jonestown in 1978 when 918 cult members took their own lives or were murdered.[8]

Perhaps the most well-publicized current example of intransigence is found among jihadists. In most instances, their beliefs have been acquired through indoctrination. Alternative beliefs are rejected. Individuals who embrace Christianity, Judaism, Hinduism, or atheism, or those who are advocates of secular societies are viewed as enemies of Islam who must be converted or killed—in effect, those who don't believe or behave properly are expendable.[9] Jihadists embrace a belief system that asserts itself as the *single truth*, tolerates minimal variance of believer's behavior, and justifies rewards without guilt for violence toward nonbelievers.[10] Card-carrying Catholics and Mormons are similar in certain ways. Often their beliefs are intransigent. But what is indoctrinated differs in one critical way: killing those who harbor alternative beliefs is rarely part of their message. Brain systems are very much involved in the process. For jihadists as well as others with intransigent beliefs, brain reading of nonbelievers' brain states serves to confirm their already-present negative attributions toward nonbelievers—in effect, one sees what one believes. Positive brain readings take place among those who share one's beliefs. Stories and models affirm what is believed—both the Koran and the Bible are striking examples of story affirmation.

Milder forms of Islam are often imagined to be tolerant of other ideologies. But in the estimate of a number of scholars, this is largely myth based on wishful thinking.[11] Perhaps this is why Chancellor Angela Merkel has recently noted that Germany's decade-long attempt to establish a civil multiculturalism in a country that is predominantly Christian with secular-based laws has failed.[12]

Jihadists' beliefs are but one example of intransigence. Intelligent design, when it is combined with a rejection of evolutionary theory, is another example: it asserts a belief and identifies the enemy of the belief. Intransigence can prevail when universities deny jobs to teachers because of their religious beliefs or their sex.[13] It can occur when people feel powerful and new opinions are ignored.[14] The list does not end here. *Similar intransigence is found among people who embrace particular lifestyles, politically correct thinking, animal rights, gender rights, and environmentalism.*[15] For these individuals, their beliefs are morally superior. There is a sense that they possess special knowledge and only they know what needs to be done for a larger good, which they alone define. Emotionally there is only one right way to feel. Cognitively there is only one right way to think. Behaviorally there is only one right way to act. The *divide*-reducing and desirable physiological effects of this type of believing have been noted.

This may seem to be a harsh evaluation of our nature. Perhaps it is.

Hopefully it is wrong. Yet it is one that evidence favors: much of humanity seems willing—perhaps even motivated—to build belief superstructures and illusions of understanding, hide from the reality of others and evidence, and experience satisfaction in doing so.[16]

The preceding points add up this way. The brain is prepared for indoctrination. Once it is achieved, contexts, symbols, and behavior trigger beliefs about how the world is supposed to work, identify the deficiencies and dangers of alternative beliefs, and imply or specify expected actions among believers. Those who indoctrinate surely understand this. They tie beliefs to personal satisfaction and pleasure, assert that believers will have access to special belief-related truths and, possibly, rewards, and affirm the moral uniqueness and superiority of their teachings. Negative attributions identify outsiders and their beliefs.

The literature is filled with reports critical of the indoctrination of children, prisoners of war, and members of ethnic groups and cults. Thus it is easy to abhor the process. Yet many of us attempt to indoctrinate our children on matters of personal safety, hygiene, manners, and open-mindedness—we just give these efforts another name, such as "responsible parenting." This is a special type of indoctrination, however. Indoctrination leading to intransigent beliefs can be viewed as a form of censorship in that alternative beliefs are dismissed or become the targets of action. In contrast, good parenting acknowledges alternatives such as teaching a child to understand and respect the lifestyles of others.

Still, things are not completely tidy. For example, there are studies documenting that indoctrination is an effective form of behavioral modification for children with serious attention deficit/hyperactivity disorder (ADHD).[17] Religious or lifestyle indoctrination is associated with improved personal health and increased longevity (see chapter 8). And what is to be made of organizations like the Central Intelligence Agency, the State Department, or highly competitive companies, which often indoctrinate their employees regarding the safeguarding of classified information and its unauthorized dissemination?

An intriguing feature of indoctrination is the frequency with which key beliefs and rituals are repeated. Consider how unusual this is. If every day of your life you are told over and over that one atom of carbon and two atoms of oxygen combine to make carbon dioxide or that Rome is closer to the equator than Paris, it would be irritating. Why then is there far less irritation when indoctrinated beliefs and rituals are repeated daily? One answer has been mentioned: there are desirable physiologic and psychological effects associated with such beliefs and behavior. Another is that repetition, such as frequent prayer, asserts one's commitment to a belief. This too is likely to have desirable physiologic and psychological effects as well as assure one's continued membership among groups whose members share the same belief.

There is another possible reason. Repetition may be essential to offset the anti-indoctrination effects of a wandering brain, situational cognition, and tensions resulting from occasional doubt about what is believed and what seems believable. Repetition suggests that intransigent beliefs are vulnerable to modification or rejection. That this can happen is suggested by the number of individuals who exit from religions, cults, and other belief-dominated groups.[18] Belief repetition, ritual, and programmed socialization may serve to reduce exit frequency. Stated another way, tasks involved in sustaining an indoctrinated belief include controlling information—perhaps also physical contexts—so that questions are not raised and experience is consistent with what is believed.

It is worth noting that behavior analogous to intransigent beliefs is observed among nonhuman primates when animals in one group treat members of other groups with the same suspicion as do many of their human cousins with other members of their species.[19] Similar behavior among humans thus may be strongly predisposed, which implies that it will be very resistant to change.

SELF-INDOCTRINATION

People self-indoctrinate and the brain assists in the process.

People seek emotional satisfaction and meaning in their lives. They desire comfort in the face of a world that they often perceive as insensitive to their needs and values. At times, they indoctrinate themselves into religions, cults, and lifestyles to achieve these ends. This was the choice of the interviewee in chapter 3 who joined the United States Marines and then a cult in his search to "be someone."

People who self-indoctrinate actively seek to assimilate dogma. Doing so is not considered an imposition, an affront, a limitation on their freedom to believe as they please, or a form of censorship. Self-indoctrinated beliefs are associated with a positive self-image, which assures that a believer's emotions and physiology are involved. The elimination of ambiguity and *divide* reduction simplify the process. When known others share the same or highly similar beliefs, one is a member of a community of believers.

CULTURAL INDOCTRINATION

There are also forms of cultural indoctrination. The cultural myths discussed in chapter 5 are examples. This type of indoctrination often suppresses existing beliefs and evidence that then become lost to members of a culture often to be rediscovered or reinvented by later generations. A familiar example is the abandonment of key parts of the Greek intellectual tradition during the reign of the Roman emperor Constantine and the low Middle Ages, followed by their subsequent rediscovery during the Renaissance and the Enlightenment.[20]

Although culturally indoctrinated beliefs tend to change slowly, there are exceptions. This was the case, for example, following the Japanese attack on Pearl Harbor on December 7, 1941. Overnight, a large percentage of Americans came to believe that the Japanese were evil and duplicitous. These beliefs persisted throughout World War II and for years after. The events of 9/11 and their association with al-Qaeda have had a similar effect for many Americans.

EXPERIENCE

Experience may lead to intransigent beliefs. A portion of an interview with a young woman associated with PETA provides a relevant example.

> *Author:* "What drew you to PETA?"
> *Interviewee:* "I was working in a medical research lab. I was very enthusiastic about the job, thought the research might help

people who were mentally ill. But after a while, I was struck by the way the research team treated animals."

Author: "For example?"

Interviewee: "The animals were young monkeys. They seemed very human to me. They all had names and recognized us. They were happy to see us each day. They became our friends, just like people. Then the experiments started. The monkeys received experimental drugs. Electrodes were placed into their brains to measure the effects. I hurt every time I saw that happen. They were never the same after the experiments. We damaged them for life."

Author: "Go on."

Interviewee: "I complained to the research director. It seemed to me that if the experiments had to be done, it would be better to sacrifice the animals after the experiments were finished rather than leave them damaged to live out their lives. I told the director so. He rejected my suggestion."

Author: "Did he tell you why?"

Interviewee: "He claimed that it was important scientifically to know the long-term effects of the drugs."

Author: "And?"

Interviewee: "I stayed around for almost a year to see the long-term effects. They were as you might expect: the monkeys were abnormal, and that didn't change. Even the director agreed. So I raised the issue of sacrifice again. Again he rejected it, this time on the basis of funding."

Author: "Funding?"

Interviewee: "Yes. Apparently there was a per diem payment from a grant for each day an animal was alive. So now the animals were being kept alive just for money. It was about then that I called some friends who worked with PETA."

Author: "And?"

Interviewee: "The rest is history. I became an active PETA member and participated in some of its activities. I was arrested. A fine and a probationary period followed."

Author: "And what is your situation now?"

Interviewee: "If you are asking whether I am an active PETA member, I won't say. But my sentiments haven't changed. There are literally hundreds of examples of animal abuse in medical and drug company laboratories. Members of PETA care and want change. Not many others care. The medical researchers and the drug companies are the worst of the bunch. They'll say anything to justify their behavior. They lie. Their research and funding are more important than caring for animals. They're even worse than slaughter houses—at least there they kill the animals and don't extend their suffering."

Author: "Has your distress lessened with time?"

Interviewee: "No. Every time I hear of an example of animal abuse, I pain."

The events experienced by the interviewee above differ little from similar reports by people whose dissatisfactions precede their joining groups, adopting certain beliefs, and, at times, engaging in illegal or violent behavior. Typically there is a series of disquieting events. These are followed by attempts to change individuals or groups that are perceived to be responsible for the events. Rarely does change occur. The sense that little will change in the future follows. Illegal or violent behavior may loom as a justifiable or necessary response.

IMAGININGS THAT BECOME BELIEFS

There are over 4,200 currently identifiable different faith groups or religions. There are over 2,000 Christian groups each harboring identifiably different beliefs. There are over 1,000 known political-ideological beliefs. There are millions of people who believe in ghosts, haunted houses, and conspiracy theories. Add to these numbers the thousands if not millions of private beliefs that are not recorded and often not shared with others. Many of these are intransigent. Very likely, many have their origins in imaginings, particularly those for which there is no foolproof disconfirming evidence.

A portion of an interview with a forty-five-year-old man that came to believe that his dog was a reincarnation of his grandfather illustrates the shift from imagination to belief.

Author: "Let me be clear about this. Are you saying that you believe your dog is the reincarnation of your grandfather?"

Interviewee: "Yes."

Author: "Can we give your grandfather and your dog names?"

Interviewee: "Sure. Mack and Spot."

Author: "Good. Can you now explain how you came to this belief?"

Interviewee: "The dog—I mean, Spot—was about a year old when the thought occurred to me. My grandfather Mack was a very loving man. He thought the world of me. Most of the time, I couldn't do anything wrong. Well, that's wasn't always true. When I did something that he didn't think was wise, he would vigorously scold me and was less loving for a time. Anyway, one day, it occurred to me that Spot acted the same way. He is very loving and my constant companion. We get along beautifully. But once in a while, I will do something that irritates him. He will bark and then sulk—just like Mack."

Author: "Are there examples of things you did that bothered Spot?"

Interviewee: "Yes. And it was really weird at first. One day I was splitting wood and left some pieces on the ground—didn't put them on the woodpile as I usually do. He began to bark. He picked up a piece of wood and took it to the pile—Mack would have done the same thing. At the time, I had other things to do, so I didn't put the wood on the pile. Spot sulked for the rest of the day."

Author: "Rather amazing, don't you think?"

Interviewee: "Yes, truly."

Author: "Might Spot's behavior be explained by coincidence or maybe he was trying to help or imitate you?"

Interviewee: "Perhaps. But there were other incidents."

Author: "For example?"

Interviewee: "One day, I was trimming trees with a chainsaw. Normally I turn off the saw while I pick up cuttings. But on that day, I left the saw running on the ground while I was piling branches. Spot started barking uncontrollably. He did so until I turned off the saw. That is exactly what Mack would have done. From that moment on, I was convinced.

The interviewee was married, the father of four children, CEO of a successful company, a respected member of his community, and not associated with a religion or a cult.

BELIEF-DISCONFIRMATION FAILURE

Belief-disconfirmation failure and intransigent beliefs go hand in hand. Failure occurs because of a person's refusal or inability to widen *divides*—"change one's mind," in everyday parlance. The brain is closed to new beliefs and evidence and their possible interpretation.

There have been many explanations of disconfirmation failures. It may be that we are stuck with ancient, irrational, stubborn brains.[21] It may be that the preference for certain beliefs and ideologies is affected by one's physiological makeup. This has been suggested regarding political preferences.[22] Or it may be that there is what has been called "the curse of knowledge": one's knowledge—think "belief"—about an event's outcome can compromise one's ability to reason about another person's beliefs.[23]

The earlier discussion of the brain's energy conservation (in chapter 9) is another possibility. The brain acts very rapidly at the neuron level. Information travels from one part of the brain to other parts in milliseconds. Yet, as noted, by comparison, the brain is strikingly slow in thinking through complex issues with contingencies. This is due to the time and energy costs involved in evaluating evidence and creating models. Energy conservation also appears to be responsible for simplifying complex information, preferring the use of available tool kits rather than occasionally approaching familiar tasks anew, decreasing ambiguity and uncertainty by narrowing *divides*, and packaging complex and often-contradictory evidence as beliefs to offset the costs associated with thinking during moments of information overload or complexity. Although establishing or creating a belief may require the expenditure of energy, once beliefs are established, the cost of believing appears to be minimal. (Energy conservation may explain findings such as boosting gamma brainwaves improves subjects' performance on abstract-reasoning tests and electrical stimulation to the brain can increase learning.[24]) In effect, changing a belief is not an energy-free activity such as correcting a misspelled word: beliefs are networked to other beliefs in the brain, to pleasure and reward centers, and so forth. Altering these many relationships is highly energy dependent and time consuming.

Combining the preceding points, *if*

- *we are born to believe,*
- *the brain is prepared to believe,*
- *people overvalue their beliefs,*
- *beliefs are associated with self-righteousness, pleasure, and reward,*
- *the brain is biased in favor of divide reduction,*
- *the brain has numerous systems that facilitate belief development and perpetuation,*
- *divide distance does not predict the strength or believability of belief,*
- *we see what we believe,*
- *emotions often determine what is believed,*
- *beliefs reduce ambiguity and uncertainty,*
- *and beliefs conserve brain energy,*

then we are endowed with a brain that will favor belief creation and disconfirmation failure and is one step ahead of consciousness.

17
AND WHY?

When it comes to ourselves, we like to think that we are open-minded and reflective. At moments, we are. There are other moments, however. Beliefs dealing with religion, politics, morals, family, ethnic groups, aesthetics, neighbors, sports heroes, science, politics and politicians, atypical behavior and desires, local and national governments, international institutions, land use, spouses, parents, children, and even pets are often devoid of even the slightest hint that they are treated by open and reflective brains. Further, the range and depth of experience and knowledge in daily life differs from person to person, place to place, and brain to brain. Different brains, places, and experiences assure that there will be a wide range of beliefs and *divides*. This is the world in which we live. There is no escape. It is deeply infused with intransigent beliefs.

Not everyone will agree with this assessment. There are optimists who maintain that the age of faith has been succeeded by a secularist age of reason, enlightenment, scientific enrichment, political liberty, and open-mindedness. Yet the scorecard of daily life reads otherwise. The contention that the world is now living in a period of scientific enrichment, enlightenment, and political liberty may apply to a small percentage of the world's population. But to contend that this state of affairs is worldwide ignores events and trends of both the last and the present century:[1] national, ethnic, and religious wars and murder; oppressive dictatorships; human exploitation; poverty; discrimination; rampant crime; and environmental destruction hardly invite the conclusion that daily life has been graced in the way optimists claim or wish.

Scientific research over the last one hundred years has clarified much about the brain and its systems that contribute to beliefs and *divides*. What is often overlooked is that there are no indications that these systems will change in the foreseeable future—evolution moves slower than a snail's pace. Moreover, each new generation begins its journey in life with a brain that is the product of millions of years of evolution. Like our arms and legs, the same brains that belonged to our great-great-grandparents will be around for generations to come.

Our brains will develop technological innovations and devise new forms of creativity. At times they will support humanitarian acts and efforts to eliminate disease, poverty, and discrimination. They will endorse efforts to address climate change and species extinction. Yet, at times, the same brains will indoctrinate others, embrace intransigent beliefs, fail to disconfirm unsupportable beliefs, resist belief change, and favor believing over skepticism and *divide* reduction over the careful review of evidence.

The probable outcome will be ongoing personal and group conflict, the emergence of new self-interest groups, and the failure to address critical questions adequately. For example, will there be enough food and water in the year 2045 when the world's population is expected to reach nine billion people? If not, will it be our lot to live out the tragedy of the commons? And how can neighbors, members of city councils, and friends who embrace diametrically opposed beliefs figure out how to get along?

Critical to an understanding of daily life is an understanding of how the brain believes and manages *divides*. It is not that this understanding will suddenly resolve many challenges and problems. What it does mean is that understanding should not be ignored. If there is clear and compelling evidence that the brain works in certain ways, then it follows that efforts to resolve problems fired by conflicting beliefs should take into account these workings. If nothing else, this accounting will partially clarify why many problems persist and remain unsolved.

Given the preceding perspective, many current factors that influence and are influenced by belief and *divides* might be discussed. Three have been selected for comment: information overload, belief fragmentation, and the time-compact present.

INFORMATION OVERLOAD

A frequently voiced explanation for the multiplicity of beliefs—"belief explosion" might be more apt—is *information overload*. Television, radio, newspapers, magazines, the Internet, e-mail, iPods, BlackBerrys, Facebook, Twitter, texting, and other forms of communication now bombard people with information from birth to death. We are inescapably entwined in an era of unparalleled information expansion.

Much information that finds its way to our brains seems unimportant. Most of the world's population has little interest in news that warlords in some distant country are once again engaged in territorial disputes, that college X has won the NCAA wrestling title for the fourth consecutive year, or that a small town in Alabama is the home of three National Merit Scholars. In addition to news items, there is an endless string of conflicting pronouncements on health, law, migration, crime, politics, poverty, economics, discrimination, and so forth.[2]

This state of affairs was slow in coming. The onset of the information age can be traced to the invention of the printing press in 1578. The subsequent development of the telegraph and short-wave radio ratcheted up both the speed of information transmission and its volume. Volume and transmission increased severalfold during the twentieth century. One consequence is that information now accumulates faster than it can be absorbed and evaluated.[3] Moreover, today, literally every technical means of communication is filled continually with information on literally every imaginable topic, a high percentage of which includes beliefs. These channels have listeners: for example, the average adult American watches TV some thirty-two hours per week and Facebook users are reported to access their own Facebook page seven times a day. The rapid spread and popularity of information of all types coupled with the time people spend accessing it invites the speculation that acquiring information is addictive. Ever more information is desired.

One result of the information explosion is that people with the same belief can select separate bits of evidence to support a belief they seemingly share. Conversely, people with very different beliefs can select the same evidence to support their different beliefs—for example, a statement by the pope can spark antithetic views among Christians, atheists, and Muslims.

In short, we live in an information world composed of an immense and ever-expanding grab bag of news, opinions, and evidence from which people select what best suits their beliefs. Selection is particularly common among people with strongly held political, religious, and moral beliefs. It is also apparent among individuals working in public media, many of whom configure their interpretation of information in ways that are consistent with their beliefs[4]—it is not clear if they can prevent themselves from doing so. The upshot is this: the number of beliefs that are associated with discordant, irrelevant, and incomplete evidence increases daily.

For many who read this, there is little that is new in the preceding paragraphs. The information explosion and its implications have been discussed widely. Some commentators foresee dire consequences. For example, the Internet has been characterized as a threat to culture because it simplifies and feeds people with bits and pieces of information.[5] As it has emerged as a dominant means of accessing information, studies show that recall of the accessed information declines while the recall of its sources increases.[6] Others have come to view the Internet, cell phones, and other electronic devices as facilitators of conflict because of their use by criminals and terrorists. Still others have predicted "reputation bankruptcy" because a person's past is difficult to erase in the digital age of data storage—Have you checked your credit score lately?[7] Around the corner, there are those who view the explosion positively.[8] They foresee easy access to information, quick and cost-efficient communication, and a broadening of awareness about important issues such as financial stability. Could this be due to neural systems that mediate optimism bias?[9]

Same information; different beliefs.

Still, on balance, a "dumbing down" of understanding is the highly likely consequence of overload. The greater the amount of available information, the less it's susceptible to in-depth analysis and the greater the superficiality of its interpretation. This is the fate of much scientific reporting. Because of its often highly technical nature, the language of science differs from that of daily life.[10] As a result, there is a tendency among writers to simplify findings and ideas, which has unnerving implications when such information finds its way to policy makers.[11]

It should not come as a surprise that the explosion has stimulated efforts at brain manipulation. For example, Google has developed algorithms that are designed to profile persons who query through their system. Once the profile is identified, responses to queries are organized according to what Google's algorithm deems will be of greatest interest to the person accessing its system.[12] The effort may be well intended. But an obvious effect is to direct a person making a Google inquiry to material that likely will reinforce the person's beliefs.[13]

In 1964 Marshall McLuhan pointed out how media affects the patterning of human associations. The same point applies to how people communicate one to another. Comparing 1964 with today, there is an inverse relationship between the rise of electronic communication and face-to-face communication. There are some desirable effects of this change. Remotely located individuals can communicate rapidly and with ease via electronic means. It's also possible (but unlikely) that, in discussing strongly held beliefs, Internet communications are more focused, less emotional, and less confrontational. Unlike what often characterizes face-to-face interactions, Internet messages can be carefully crafted and edited. But there is also information that is lost, such as a shared context and the opportunity to observe and interpret nonverbal behavior, voice pitch, and brain read. The trend of less and less face-to-face dialogue raises the possibility of an increase in intransigent beliefs due to reduced opportunities for belief modification.

BELIEF FRAGMENTATION

Belief fragmentation is not new.[14] It comes and goes. In its current version, it is largely a consequence of information overload, the excessive number of beliefs that overload fosters, and the way the brain works. In its extreme form, no two people would share even similar beliefs—each of us would live in our own private world of understanding and meaning.

Cultural and religious myths and their political or moral equivalents return to the discussion at this point. Without the influence of group-uniting and shared myths or belief systems, individuals tend to believe whatever suits them. The more they do so, the greater the degree of fragmentation.

The American myth provides an instructive illustration. For almost four hundred years, the King James Bible provided a framework for beliefs, values, expressive symbols, and artistic motifs in which individuals—not all, of course, but many—defined and interpreted their social world and behavior and made judgments. Interpretations of the Bible served as a reference point for assessing behavior and decision making. The belief system provided non-legal guidelines for perhaps 90 percent of our behavior—that is, for behavior not covered by formal laws.[15] Without such guidelines, communities break down. Social participation declines. Personal and legal conflicts increase. Individual disenfranchisement follows. Consensus-based morality tends to disappear.[16] Chronic stress may follow and, when it does, it is often accompanied by a biasing of behavior toward habit, which includes beliefs.[17] These are outcomes that the advocates of secularism didn't foresee.

To digress a moment to offer an analogy, what happens when beliefs fragment and there is no overarching belief system that people embrace to guide their behavior would be much like a football game in which all the players played by their own private rules.

Let me be clear about the preceding paragraphs. The point here is not that of passing judgment on whether one cultural myth is superior to others (although I have a view) or about the pros and cons of secularism. Rather the point is that a shared belief system is essential for social cohesion, otherwise beliefs fragment and social cohesion declines. The less a myth is shared by individuals, the greater the disruption of daily life. Belief polarization and self-righteousness about one's beliefs can quickly follow. Discounting the merit of others and their beliefs takes on a reflex-like nature. As an example, take American politics from approximately the 1970s to the present. It is a period characterized by increasing character assassination of politically elected and aspiring officials often without a shred of evidence.

There are disturbing downsides to these points. Periods of fragmentation invite conflict of all types, including interpersonal, aesthetic, national, international, military, and legal. Fragmentation chokes off conflicting evidence and fosters *divide* reduction as a way of managing information and reducing energy expenditure by the brain. Believers come to attribute absolute certainty to their beliefs irrespective of their plausibility or evidence—for example, concerned individuals mount serious political campaigns against "smart electric meters" because they believe that the meters cause cancer. Doubts about one's beliefs tend to disappear. Individuals deny the presence of global warming and dismiss evidence of human-caused environmental destruction and species extinction. Claims are made that those who believe in God are delusional.[18] The net effect is turbulent times, turbulent brains, and turbulent emotions. In turn, there is an opportunity for perverse and destructive beliefs as well as oppressive doctrines to exert their influence.

As beliefs become personalized, attributions follow suit and serve as proxies for belief-supporting evidence. Similar scenarios apply to the readings of others' brain states and triggering—for example, simply seeing a public official or a next-door neighbor may initiate a strong positive or negative response. Stories and models accommodate to these responses.

TIME-COMPACT PRESENT

It is Friedrich Nietzsche to whom the following statement is widely attributed: "Modern man eats knowledge without hunger." What he might have meant is that modern man absorbs information without passion, necessity, and analysis. To do so is a trademark of the time-compact present.

In 1999 J. T. Fraser coined the term *time-compact present*: the present—now!—becomes the dominant focus of attention and emotional investment.[19] There is an excessive emphasis on the now at the expense of preserving what is valuable from the past. The uncertainty of the future disappears and is replaced by imaginings of infinite options. In effect, the present is cognitively and emotionally decoupled from the past and the future. Stories and models that depend on the past or a likely future to gain their structure and meaning in turn lose their relevance. What happened yesterday doesn't mean that it will happen today or tomorrow. Information overload, brain reading, and attribution hype the here and now. It's as if, as a species, we have rejected the wisdom of history and a serious appreciation of the uncertainty of moments to come.

Time-compactness means that the half-life of information—even very important information—is reduced sharply. For example, recall the March 29, 2011, Tohoku-Oki earthquake in the sea off of Japan.[20] For several days, reports about human suffering, physical damage, rising radiation levels, and a seriously compromised nuclear power plant dominated the news. Except largely for Japan, by August 1, 2012, news about the earthquake and its details had all but disappeared. The event had begun its trip into the forgotten past.

With the loss of a time-related perspective, unrealistic hopes surge—yet another time-compact trademark. For example, during the past few decades, there have been repeated attempts to improve American education. In some schools, this has happened. But the overall goal of improvement is difficult to reconcile with high school graduation rates that range from 58 percent to 93 percent, as well as a breakdown in general literacy.[21] Over the same period, universities have become largely fractious collections of interest groups.[22] Progressively they have scrapped their traditional aim of graduating informed students in favor of preparing them for jobs or indoctrinating them with this or that ideology.[23] This situation exists in part because of the consequences of time-compactness: imaginings and beliefs trump facts and an appreciation of what can be learned from the past and reasonably expected in the future.

History however has its own timetable largely irrespective of what the brain and culture are up to. The year 1908 saw the dawn of flight, a competitive but nonhostile race to the South Pole, and the invention of the Model T Ford.[24] Over a century later, the ramifications of these events have yet to fully play out. Satellites fly around the globe, collecting information for both good and bad reasons. The world is clogged with automobiles and their combustion products. And nations spar over the control of Antarctica.

Or consider the disappearance of a future for much of the earth's natural environment and animal populations. Despite serious and expensive efforts to preserve endangered species and isolated pockets of progress, the battle is being lost, so much so that a number of scientists have suggested that the Earth's sixth mass extinction is already underway.[25] *It is as if, as a species, we are helpless to address these issues successfully.* Intransigent beliefs, information overload, and belief fragmentation are three of the reasons why. The time-compact present is a fourth.

18
WHAT TO DO?

There is a vaudeville act that goes something like this:

The stage is dark except for a streetlight that illuminates an area a few feet from its base. A man is searching for something in the light.

A policeman arrives and asks, "What are you looking for?"

"My keys," the man replies.

The policeman joins in the search for several minutes but fails to find the keys. Frustrated, he turns to the man, points to the lighted area and asks, "Are you sure you lost your keys there?"

"Not there," replies the man.

"Where then?" asks the policeman.

The man points to the dark.

"Then why are you searching here?" asks the policeman.

"The light's better," replies the man.

The original intent of this book was not to develop suggestions dealing with possible ways of changing beliefs and *divides*. Rather, my aim was to write about the amazing features of the brain that contribute to them. As the book has evolved, it is possible to infer some suggestions. I will come to these below. First, a few words about context.

Our brains aren't going to change in the foreseeable future. Our strong tendency to believe isn't going to disappear. The nature of human nature isn't going to undergo a miraculous transformation overnight. Motivations to succeed, control others, obtain fame, and act in self-interested ways aren't going to go away. Granting these points, it is easy to imagine why people seek solutions to problems in the light. New laws, regulations, and ideological persuasion are typical means of doing so. Yet these have proven successful on only occasion. The presence of intransigent beliefs and belief fragmentation means that there will be significant resistance, often open opposition, to any suggested solutions, whatever their merit. For example, a June 25, 2012, edition of the *Wall Street Journal* featured a section titled "Squaring Off on Education" in which it reported on responses of experts to six questions: (1) Should all US students meet a single set of national proficiency standards? (2) Should student test scores be used to evaluate teachers? (3) Should more college financial aid be based on need, not merit? (4) Do too many young people go to college? (5) Should tenure for college professors be abolished? (6) Should colleges consider legacies in the admissions process? Sturdy "yes" and "no" answers were offered for each question.

Education is often proposed as a solution to the perceived ills of society. Thus it is not surprising that recently there has been an increased call for the teaching of science and mathematics at all levels of education.[1] These well-intended recommendations are based on the belief and hope that science and math education will transform ways of thinking and, among other things, significantly reduce the prevalence of unfounded and intransigent beliefs and their consequences.[2] This hope is closer to an illusion than reality, however. There is a limited number of beliefs that science and math can address. *Most of the world runs most of the time on beliefs that science can't address. They reside in the dark, not in the light that science might shed.*

Now to the suggestions.

First, be skeptical about everything you and others believe—this book is no exception. There is nothing in the dark about this suggestion. It has been around since the Enlightenment. At times, skepticism is easy, as in instances when others express beliefs with which one disagrees. But when it comes to our own beliefs, skepticism can be very hard. The brain has assured that.

Second, intensify the teaching of a liberal education—that is, an education that introduces students to different models, teaches strong analytic skills, exposes them to how the brain works, and respects both the past and the uncertainty of the future. Currently this suggestion is in the dark. Why this type of education? Because models, analytic skills, knowledge of the past, and details about how the brain works are the best safeguards we have to offset the effects of the time-compact present, beliefs without supporting evidence, intransigent beliefs, and belief disconfirmation failure.

Third, remember that belief representations in the unperceived brain are composed of networks of information created from a variety of internal and external sources and systems, including triggering, brain reading, imaginings, stories, models, direct and indirect evidence, inferences, and far more. The brain is prepared for belief creation and acceptance. It is biased in favor of *divide* reduction. It prefers beliefs that are pleasurable and rewarding to those that are unpleasant and aversive. It is composed of systems that organize and process information in ways over which we have minimal influence. Evidence has only limited efficacy in altering beliefs. These many factors assure that for the foreseeable future there will be a near-endless number of beliefs and *divides*. Intransigent beliefs will contribute prominently to this number.

✳✳✳

Some Updates

As scheduled, Mrs. X and I met several months after we had temporarily parted ways. Her belief that she was not the child of her parents had disappeared. Why and how remains a mystery.

Greg returned to the United States, married Francesca, and turned his attention to history.

Howard remained the rogue psychologist.

The vervet monkeys on Saint Kitts continue to thrive.

And finally. When I was a boy, I believed in ghosts. Then for decades I didn't. Now I wonder: there may be belief representations of ghosts located somewhere in my brain.

NOTES

CHAPTER 2. WHERE TO START?

1. *Merriam-Webster's Collegiate Dictionary*, 11th ed. (Springfield, MA: Merriam-Webster, 2003), s.v. "belief."

2. R. R. Britt, "Monsters, Ghosts, and Gods: Why We Believe," *LiveScience*, August 18, 2008.

3. M. Alvarez, *Kinds of Reasons* (Oxford: Oxford University Press, 2010); see also J. O. Beahrs, "Self-Deception and Intrapsychic Structure," *American Psychiatric Association*, May 17, 1990.

4. L. B. Steadman et al., "Toward a Testable Definition of Religious Behavior," in *The Biology of Religious Behavior*, ed. J. Feierman (Santa Barbara, CA: Praeger, 2009), pp. 20–35.

5. S. Bowles and H. Gintis, *A Cooperative Species* (Princeton, NJ: Princeton University Press, 2011).

6. J. Feierman, personal communication with the author.

CHAPTER 3. TYPES AND USES

1. C. MacKay, *Extraordinary Popular Delusions and the Madness of Crowds* (1841; repr., Lexington, KY: Maestro Reprints, 2010).

2. A. Mazur, *Implausible Beliefs* (New Brunswick, NJ: Transaction Publishers, 2008).

3. L. Tiger, *Men in Groups* (New York: Random House, 1969); see also M. Alvarez, *Kinds of Reasons* (Oxford: Oxford University Press, 2010).

4. J. Denrell, "Indirect Social Influence," *Science* 321 (2008): 47–48.

5. R. V. Hine, *California's Utopian Colonies* (1953; repr., Berkeley: University of California Press, 1983); see also R. A. Billington, *The Far Western Frontier* (New York: Harper and Row, 1956); H. N. Smith, *Virgin Land* (1950; repr., Cambridge, MA: Harvard University Press, 1971); and W. H. Goetzmann and W. N. Goetzmann, *The West of the Imagination* (New York: Norton, 1986).

6. E. Shackleton, *South* (1919; repr., New York: Lyons Press, 1998); see also F. A. Worsley, *Shackleton's Boat Journey* (New York: Norton, 1977).

7. T. Jeal, *Stanley* (New Haven, CT: Yale University Press, 2007).

8. J. Hayman, ed., *Sir Richard Burton's Travels in Arabia and Africa* (San Marino, CA: Huntington Library, 2005).

9. M. C. Keller and G. Miller, "Resolving the Paradox of Common, Harmful, Heritable Mental Disorders: Which Evolutionary Genetic Models Work Best?" *Behavioral and Brain Sciences* 29 (2006): 385–452.

10. H. B. Stowe, *Uncle Tom's Cabin* (1852).

11. P. Singer, *Animal Liberation* (New York: Random House, 1975).

12. J. S. Foer, *Eating Animals* (London: Penguin, 2010).

13. P. Bloom, "How Do Morals Change?" *Nature* 464 (2010): 490.

14. L. Tiger and M. McGuire, *God's Brain* (Amherst, NY: Prometheus Books, 2010).

15. C. Burnes, *Deadly Decisions* (Amherst, NY: Prometheus Books, 2008).

16. W. Northcutt, *The Darwin Awards* (New York: Dutton, 2000).

17. J. Bohannon, "The Nile Delta's Sinking Future," *Science* 327 (2010): 1444–47.

18. E. Quill, "Can You Hear Me Now?" *Science News*, April 24, 2010.

19. "List of Philosophies," *Wikipedia*, last modified May 27, 2013, http://en.wikipedia.org/wiki/List_of_philosophies (accessed May 29, 2013).

20. "University of Oregon Survey of Beliefs and Opinions," http://darkwing.uoregon.edu/~prsnlty/surveySBO/SBOlist.htm (accessed May 3, 2010).

21. A. M. Josephy Jr., *500 Nations* (New York: Knopf, 1994).

22. "Unusual Trivia Collection," http://www.corsinct.com/trivia/scary/html (accessed May 3, 2010).

CHAPTER 4. WHAT PSYCHOLOGISTS HAVE FOUND

1. T. Gilovich, *How We Know What Isn't So* (New York: Free Press, 1991).

2. M. Shermer, *Why People Believe Weird Things* (New York: St. Martin's Griffin, 1997), p. 297.

3. Ibid.

4. T. Kida, *Don't Believe Everything You Think* (Amherst, NY: Prometheus Books, 2006).

5. C. A. Fine, *A Mind of Its Own* (New York: Norton, 2006).

6. M. Shermer, *The Believing Brain* (New York: Times Books, 2011).

7. D. Kahneman, *Thinking Fast and Slow* (New York: Farrar, 2011).

8. D. Eagleman, *Incognito* (New York: Pantheon, 2011).

CHAPTER 5. LESSONS FROM HISTORY

1. J. G. Frazer, *The New Golden Bough* (1890; repr., New York: Criterion Books, 1959); see also J. Barzun, *From Dawn to Decadence* (New York: HarperCollins, 2000); C. Freeman, *The Closing of the Western Mind* (New York: Vintage, 2005); F. K. Salter, ed., *Welfare, Ethnicity, and Altruism* (London: Frank Cass, 2004); T. Kamusella, *The Politics of Language and Nationalism in Modern Central Europe* (London: Palgrave Macmillan, 2009); and T. Bulfinch, *Bulfinch's Mythology* (1855; repr., New York: HarperCollins, 1991).

2. J. Lennon, *Irish Orientalism* (Syracuse, NY: Syracuse University Press, 2008).

3. Ibid.

4. Ibid.

5. R. G. Collingwood, *The Idea of History* (1946; repr., New York: Galaxy, 1956).

6. S. Kennedy, ed., *Beckett and Ireland* (Cambridge: Cambridge University Press, 2010).

7. Kamusella, *Politics of Language and Nationalism*.

8. M. Laruelle, *Russian Eurasianism* (Baltimore: Johns Hopkins University Press, 2008).

9. Ibid.; see also J. Weatherford, *Genghis Khan* (New York: Three Rivers Press, 2004).

10. F. O. Matthiessen, *American Renaissance* (New York: Oxford University Press, 1941); see also H. N. Smith, *Virgin Land* (1950; repr., Cambridge, MA: Harvard University Press, 1971); H. Zinn, *A People's History of the United States* (New York: HarperCollins, 1980); C. N. Degler, *Out of Our Past* (1959; repr., New York: Harper and Row, 1970); P. A. Smith, *A New Age Now Begins* (New York: McGraw-Hill, 1976); R. A. Billington, ed., *The Reinterpretation of Early American History* (San Marino, CA: Huntington Library, 1966); A. de Tocqueville, *Democracy in America*, translation of the original 1848 edition by J. P. Mayer (Garden City, NY: Doubleday, 1969); P. Miller, *The Life of the Mind in America* (New York: Harcourt Brace, 1965); W. L. Warner, *American Life* (1953; repr., Chicago: University of Chicago Press, 1962); R. M. Dorson, *American Folklore* (Chicago: University of Chicago Press, 1959); and J. Rakove, *Revolutionaries* (New York: Houghton Mifflin Harcourt, 2010).

11. W. J. Johnson, *A Dictionary of Hinduism* (Oxford: Oxford University Press, 2009).

12. E. Cameron, *Enchanted Europe* (Oxford: Oxford University Press, 2010).

13. *Merriam-Webster's Collegiate Dictionary*, 11th ed. (Springfield, MA: Merriam-Webster, 2003), s.v. "myth."

14. Ibid., s.v. "belief."

15. P. G. Maxwell-Stuart, *Satan* (London: Amberley Publishing, 2009).

16. P. Ricoeur, *The Symbolism of Evil* (Boston: Beacon Press, 1969); see also C. Sagan, *The Demon-Haunted World* (New York: Ballantine, 1997); M. Alvarez, *Kinds of Reasons* (Oxford: Oxford University Press, 2010); L. Tiger and M. McGuire, *God's Brain* (Amherst, NY: Prometheus Books, 2010); P. Ackroyd, *The English Ghost* (London: Chatto and Windus,

2010); and T. Chesters, *Ghost Stories in Late Renaissance France* (Oxford: Oxford University Press, 2011).

17. Ricoeur, *Symbolism of Evil*.

18. M. B. Norton, *In the Devil's Snare* (New York: Knopf, 2002).

19. J. C. Baroja, *The World of Witches* (1964; repr., London: Phoenix Press, 2001).

20. Ibid.

21. Ricoeur, *Symbolism of Evil*.

22. J. Feierman, personal communication with the author.

23. K. Knudsen, "Ghana Health Survey," *Medicine & Health/HIV & AIDS*, March 24, 2010.

24. S. M. McClure, "Neural Correlates of Behavioral Preference for Culturally Familiar Drinks," *Neuron* 44 (2004): 379–87; see also N. Singer, "Making Adds That Whisper to the Brain," *New York Times*, November 14, 2010.

25. E. Thompson, *Mind in Life* (Cambridge, MA: Harvard University Press, 2007).

26. L. Brothers, *Friday's Footprint* (New York: Oxford University Press, 1997).

27. D. Wigal, *Historic Maritime Maps* (London: Sirrocco, 2007).

28. Ibid.

29. D. Carrasco, ed., *Mesoamerican Cultures* (New York: Oxford University Press, 2001); see also C. C. Mann, *Ancient Americans* (London: Granta Books, 2005); M. D. R. Martinez et al., "Oldest Writing in the New World," *Science* 313 (2006): 1610–14.

30. N. Crane, *Mercator* (New York: Henry Holt, 2002).

31. D. C. Dennett, *Darwin's Dangerous Idea* (New York: Simon and Schuster, 1995).

32. E. O. Wilson, *From So Simple a Beginning* (New York: Norton, 2006).

33. T. S. Kuhn, *The Structure of Scientific Revolutions* (Chicago: University of Chicago Press, 1962).

34. J. D. Miller et al., "Public Acceptance of Evolution," *Science* 313 (2006): 765.

35. Ibid.

36. E. G. Scott, "Defending the Teaching of Evolution in the Public Schools," *National Center for Science Education* (April 2010).

37. J. Mervis, "Tennessee House Bill Opens Door to Challenges to Evolution, Climate Change," *Science* 332 (2011): 295; see also Scott, "Defending the Teaching of Evolution."

38. M. B. Berkman and E. Plutzer, "Defeating Creationism in the Courtroom, but Not in the Classroom," *Science* 331 (2011): 404–405.

39. R. Dawkins, *The Greatest Show on Earth* (London: Bantam, 2009).

40. J. Parker, *The Astrologer's Handbook* (Sebastopol, CA: CRCS Publications, 1995); see also W. F. Williams, ed., *Encyclopedia of Pseudoscience* (New York: Facts on File, 2000); W. Shumaker, *The Occult Sciences in the Renaissance* (Berkeley: University of California Press, 1972); and S. Reardon, "The Alchemical Revolution," *Science* 332 (2011): 914–15.

41. Shumaker, *Occult Sciences in the Renaissance*.

42. Ibid.; see also Reardon, "Alchemic Revolution."

43. Reardon, "Alchemic Revolution."

44. S. Shapin, *Never Pure* (Baltimore: Johns Hopkins University Press, 2010).

45. G. M. Edelman, *Neural Darwinism* (New York: Basic Books, 1987).

CHAPTER 6. EVIDENCE, SOURCES, AND INTERPRETATION

1. M. Heidegger, *Being and Time* (London: SCM Press, 1962). Originally published 1927; see also J. Derrida, *Of Sprit: Heidegger and the Question* (Chicago: University of Chicago Press, 1989); and M. Foucault, *History of Madness* (New York: Routledge, 2006).

2. E. S. Reich, "G-whizzes Disagree over Gravity," *Nature* 466 (2010): 1030; see also G. Amelino-Camelia, "Gravity's Weight on Unification," *Nature* 468 (2010): 40–41; R. Davis, "Big G Revisited," *Nature* 468 (2010): 181–83; and T. Siegfried, "A New View of Gravity," *Science News*, September 25, 2010.

3. P. Ball, "Beyond the Bond," *Nature* 469 (2011): 26–28; see also R. Ehrenberg, "Chemists Want You to Know That Atomic Weights Aren't Constant," *Science News*, January 29, 2011; and D. Clery, "Which Way to the Island?" *Science* 333 (2011): 1377–79.

4. T. H. Saey, "Scientists Still Making Entries in Human Genetic Encyclopedia," *Science News*, November 6, 2010.

5. K. Popper, *Conjectures and Refutations* (London: Routledge and Kegan Paul, 1963).

6. J. M. Ackerman et al., "Incidental Haptic Sensations Influence Social Judgments and Decisions," *Science* 328 (2010): 1712–15; see also A. Lleras, "Body Movements Can Influence Problem Solving," *Medicine & Health/Psychology & Psychiatry*, May 12, 2009.

7. S. Pepper, *World Hypotheses* (Berkeley: University of California Press, 1942).

8. C. Chabris and C. Simons, *The Invisible Gorilla* (New York: Crown 2010), pp. 7, 151.

9. Ibid.

10. C. Schmeltzer and H. Markovits, "Belief Revision, Self-Construction and Systematic Certainty," *Behavior, Brain & Cognition* 17 (2005): 1–9.

11. P. McCrone, "The Power of Belief," *New Scientist*, March 2004.

12. R. Ingpen and P. Wilkinson, *Encyclopedia of Mysterious Places* (New York: Viking, 1990).

13. D. Clery, "Taking Laser Science to the Extreme," *Science* 328 (2010): 806–807; see also R. Cowen, "Inventing the Light Fantastic," *Science News*, May 8, 2010.

14. L. Nordling, "Researchers Launch Hunt for Endangered Data," *Nature* 468 (2010): 17.

15. W. Leahy, ed., *Shakespeare and His Authors* (London: Continuum, 2010); see also J. Shapiro, *Contested Will* (London: Faber, 2010); R. Fox, *Shakespeare's Education* (Buchholz, Ger: Laugwitz Verlag, 2012).

16. P. Howlett and M. S. Morgan, eds., *How Well Do Facts Travel?* (Cambridge: Cambridge University Press, 2011).

17. Pepper, *World Hypotheses*.

18. Chabris and Simons, *Invisible Gorilla*.

19. T. Masuda, "Eastern and Western Cultures See Things Very Differently," *Evolutionary-Psychology*, March 5, 2008.

20. *Science* 328 (2010): 997.

21. R. E. Nisbett and T. Masuda, "Culture and Point of View," *Proceedings of the National Academy of Sciences* 100 (2003): 11163–70.

22. J. Haidt, "Revealing the Origins of Morality—Good and Evil, Liberal and Conservative," *Science Daily*, May 18, 2007.

23. King, 2006. Source lost.

24. C. A. Anderson, "Belief Perseverance," in *Encyclopedia of Social Psychology*, ed. R. F. Baumeister and K. D. Vohs (Thousand Oaks, CA: Sage 2007), pp. 109–10.

25. J. Bond, "Risk School," *Nature* 461 (2009): 1189–92.

26. K. Bach, "Critical Notice," review of *Perspectives on Self-Deception*, ed. Brian P. McLaughlin and Amelie Oksenberg Rorty, http://userwww.sfsu.edu/kbach/SDreview.htm (accessed June 3, 2013); see also R. Trivers, *Social Evolution* (Menlo Park, CA: Benjamin/ Cummings, 1985); R. Trivers, "The Evolution of Reciprocal Altruism," *Quarterly Review of Biology* 46 (1971): 35–57.

27. J. Ree, "All the Time," *Times Literary Supplement*, November 26, 2010.

28. C. Frost et al., "The Psychology of Self-Deception as Illustrated in Literary Characters," *Janus Head*, March 19, 2003.

29. E. Pennisi, "Talking in Tongues," *Science* 303 (2004):1321–23.

30. J. R. Searle, *Making the Social World* (Oxford: Oxford University Press, 2010).

31. H. Ledford, "Mental-Health Guide Accused of Overreach," *Nature* 479 (2011): 14; see also H. Ledford, *Nature* 279 (2012): 21.

32. E. Kintisch, "Critics Are Far Less Prominent Than Supporters," *Science* 328 (2010): 1622.

33. M. Zaho and S. W. Running, "Drought-Induced Reduction in Global Terrestrial Net Primary Production from 2000 through 2009," *Science* 329 (2010): 940–43; see also T. P. Barnett et al., "Potential Impacts of a Warming Climate on Water Availability in Snow-Dominated Regions," *Nature* 438 (2005): 303–309; J. Tollefson, "An Erosion of Trust?" *Nature* 466 (2010): 24–26; W. W. Iummerzeel et al., "Climate Change Will Affect the Asian Water Towers," *Science* 328 (2010): 1382–86; D. Fox, "Could East Antarctica Be Headed for Big Melt?" *Science* 328 (2010): 1630–31; and D. M. Sigman et al., "The Polar Ocean and Glacial Cycles in Atmospheric CO_2 Concentration," *Nature* 466 (2010): 47–55.

34. R. J. Nicholls and A. Cazenave, "Sea-Level Rise and Impact on Costal Zones," *Science* 328 (2010): 1517–22.

35. S. C. Doney, "The Growing Human Footprint on Costal and Open-Ocean Waters," *Science* 328 (2010): 1512–16; see also O. Hoegh-Guldberg and J. F. Bruno, "The Impact of Climate Change on the World's Marine Ecosystems," *Science* 328 (2010): 1523–28.

36. M. Zahn and H. von Storch, "Decreasing Frequency of North Atlantic Polar Lows Associated with Future Climate Warming," *Nature* 467 (2010): 309–12; see also M. E. Dillon et al., "Global Metabolic Impacts of Recent Climate Warming," *Nature* 467 (2010): 704–706; N. Gilbert, "Biodiversity Hope Faces Extinction," *Nature* 467 (2010): 764; M. Jung et al., "Recent Decline in the Global Land Evapotranspiration Trend Due to Limited Moisture Supply," *Nature* 467 (2010): 951–54; A. A. Lacis et al., "Atmospheric CO_2: Principal Control Knob Governing Earth's Temperature," *Science* 330 (2010): 356–59; and N. Jones, "Human Influence Comes of Age," *Nature* 473 (2011): 133.

37. R. Secord et al., "Continental Warming Preceding the Palaeocene-Eocene Thermal Maximum," *Nature* 467 (2010): 955–58; see also Kintisch, "Critics Are Far Less Prominent Than Supporters."

38. P. N. Pearson, "Increased Atmospheric CO_2 during the Middle Eocene," *Science* 330 (2010): 763–64.

39. W. F. Ruddiman, "A Paleoclimatic Enigma," *Science* 328 (2010): 838–39.

40. S. Jasanoff, "Testing Time for Climate Science," *Science* 328 (2010): 695–96; see also P. A. Stott and P. W. Thorne, "How Best to Log Local Temperatures?" *Nature* 465 (2010): 158–59; P. Kitcher, "The Climate Change Debates," *Science* 328 (2010): 1230–34; and Pearson, "Increased Atmospheric CO_2 during the Middle Eocene."

41. J. Tollefson, "A Chilly Season for Climate Crusaders," *Nature* 467 (2010c): 762–63.

42. M. Lott, "Eight Botched Environmental Forecasts," *Fox News*, December 30, 2010.

43. J. Tollefson, "Climate Talks Focus on Lesser Goals," *Nature* 468 (2010d): 488–89; see also Tollefson, "Chilly Season for Climate Crusaders"; and Lott, "Eight Botched Environmental Forecasts."

CHAPTER 7. SEEING WHAT WE BELIEVE

1. G. L. Walsh and the Bradshaw Foundation, *Bradshaws: Ancient Rock Paintings of North-West Australia* (Carouge-Geneva, Switz.: Edition Limitée, 1994).

2. G. L. Walsh, *Bradshaw Art of the Kimberley* (Toowong, Queensland, Aus.: Takarakka Nowan Kas Publications, 2010).

3. F. Bacon, *Francis Bacon: The Major Works*, ed. Brian Vickers (Oxford: Oxford University Press, 2006), p. 112.

4. C. A. Anderson, "Belief Perseverance," in *Encyclopedia of Social Psychology*, ed. R. F. Baumeister and K. D. Vohs (Thousand Oaks, CA: Sage, 2007), pp. 109–10.

5. R. E. Nisbett and L. D. Ross, *Inference: Strategies and Shortcomings of Social Judgment* (Englewood Cliffs, NJ: Prentice-Hall, 1980).

6. G. Quarton et al., "Man-Machine Natural Language Exchanges Based on Selected Features of Unrestricted Input (1): The Development of the Time-Shared Computer as a Research Tool in Studying Dyadic Communication," *Journal of Psychiatric Research* 5 (1967): 165–77; M. T. McGuire et al., "Man-Machine Natural Language Exchanges Based on Selected Features of Unrestricted Input (2): The Development of the Time-Shared Computer as a Research Tool in Studying Dyadic Communication," *Journal of Psychiatric Research* 5 (1967): 179–91.

7. M. Shermer, *The Believing Brain* (New York: Times Books, 2011).

8. M. Nissani et al., "Experimental Studies of Belief Dependence of Observations and of Resistance to Conceptual Change," *Cognition & Instruction* 9 (1992): 97–111.

9. P. Boyer, "Religion: Bound to Believe?" *Nature* 455 (2008): 1038–39.

10. R. R. Britt, "Monsters, Ghosts, and Gods: Why We Believe," *Live Science*, August 18, 2008.

11. D. L. Rosenhan, "On Being Sane in Insane Places," *Science* 179 (1973): 250–58.

CHAPTER 8. RELIGION AS AN EXCEPTION TO SCIENCE—OR IS IT?

1. Francis Aveling, *The Catholic Encyclopedia*, vol. 2 (New York: Robert Appleton, 1907), s.v. "belief," http://www.newadvent.org/cathen/02408b.htm (accessed June 4, 2013).

2. P. Bloom, "Religion Is Natural," *Developmental Science* 10 (2007): 147–51; see also P. Bloom, "Is God an Accident?" *Atlantic Monthly* (December 2005): 105–12; T. Tremlin, *Minds and Gods: The Cognitive Foundations of Religion* (New York: Oxford University Press, 2006); J. Barrett, "Humans 'Predisposed' to Believe in Gods and the Afterlife," *Other Sciences/Social Sciences*, May 16, 2011; D. C. Dennett, *Breaking the Spell: Religion as a Natural Phenomenon* (New York: Viking/Allen Lane, 2006); and P. Ricoeur, *The Symbolism of Evil* (Boston: Beacon Press, 1969).

3. S. J. Gould, *Rocks of Ages: Science and Religion in the Fullness of Life* (New York: Ballantine, 1999).

4. B. Graham, "Are You Far from Home?" *Decision* (September 2007): 4.

5. K. Armstrong, "Two Paths to the Same Old Truths," *New Scientist*, July 30, 2005.

6. J. Feierman, "Pedophilia: Its Relationship to the Homosexualities and the Roman Catholic Church, Part I," *Antonianum* 85 (2010): 451–77.

7. R. Dawkins, *The God Delusion* (New York: Bantam Books, 2006); see also Dennett, *Breaking the Spell*.

8. J. Repcheck, *Copernicus' Secret* (New York: Simon and Schuster, 2007).

9. O. Neugebauer, *The Exact Sciences in Antiquity* (1957; repr., New York: Dover, 1969).

10. K. Popper, *The Logic of Scientific Discovery* (1935; repr., London: Routledge, 2009); see also I. Lakatos, *Proofs and Refutations* (Cambridge: Cambridge University Press, 1976).

11. T. S. Kuhn, *The Structure of Scientific Revolutions* (Chicago: University of Chicago Press, 1962).

12. A. Zewail, "Curiouser and Curiouser: Managing Discovery Making," *Nature* 468 (2010): 342.

13. H. Waltzman, "Chemists Help Archaeologists to Probe Biblical History," *Nature* 468 (2010): 614–15.

14. M. D. Coe, *The Maya*, 7th ed. (New York: Thames and Hudson, 2005); see also "Stonehenge World Heritage Site Management Plan," *UNESCO*, July 18, 2008; M. Balter, "Was North Africa the Launch Pad for Modern Human Migrations?" *Science* 331 (2011): 20–23.

15. J. Lehrer, "The Truth Wares Off," *New Yorker*, December 13, 2010.

16. Gould, *Rocks of Ages*.

17. E. B. Davies, *Why Beliefs Matter: Reflections on the Nature of Science* (New York: Oxford University Press, 2010).

18. J. Jackson, *A World on Fire* (New York: Penguin, 2005).

19. A. Gann and J. Witkowski, "The Lost Correspondence of Francis Crick," *Nature* 467 (2010): 519–24.

20. S. Okasha, "Altruism Researchers Must Cooperate," *Nature* 467 (2010): 653–55.

21. R. Dalton, "Disputed Ground," *Nature* 466 (2010): 176–78.

22. H. Caton, "Truth Management in the Sciences," *Search* 19 (1988): 242–44; see also E. Callaway, "Report Finds Massive Fraud at Dutch Universities," *Nature* 479 (2011): 15; E. S. Reich, "Biologist Spared Jail for Grant Fraud," *Nature* 474 (2011): 552.

23. J. Beahrs, personal communication with the author.

24. D. Kahan, "Why Scientific Consensus Fails to Persuade," *Phys.org*, September 14, 2010.

25. Ibid.

26. C. Scheitle, "Losing Your Religion Deemed Unhealthy," *Medicine & Health/ Psychology & Psychiatry*, September 22, 2010.

27. C. Drew, "New Studies Show Reduced Depression with Transcendental Meditation," *Medicine & Health/Psychology & Psychiatry*, April 7, 2010.

28. K. J. Flannelly et al., "Beliefs about Life-after-Death, Psychiatric Symptomology and Cognitive Theories of Psychopathology," *Journal of Psychology and Theology* 36 (2008): 94–103.

29. A. Newberg and M. R. Waldman, *How God Changes Your Brain* (New York: Ballantine, 2009).

30. B. Bower, "Happiness Found in Next Pew Over," *Science News*, January 1, 2011.

31. J. Feierman, "The Image of God to Whom We Pray: An Evolutionary Psychobiological Perspective," *Pensamiento* 67254 (2012): 817–29.

32. L. Tiger and M. McGuire, *God's Brain* (Amherst, NY: Prometheus Books, 2010); see also R. A. Scott, *Miracle Cures* (Berkeley: University of California Press, 2010).

33. N. Krause, "Church-Based Social Relationships and Change in Self-Esteem over Time," *Journal for the Scientific Study of Religion* 48 (2009): 744–88.

34. P. McNamara, "The Motivational Origins of Religious Practices," *Zygon* 37 (2002): 143–60.

35. R. Sloan, quoted in J. Kluger, "The Biology of Belief," *Time*, February 12, 2009.

36. L. Wolpert, *Six Impossible Things before Breakfast* (London: Faber and Faber, 2006).

37. G. Basalla, *Civilized Life in the Universe* (New York: Oxford University Press, 2005).

38. F. Tipler, *The Physics of Christianity* (New York: Doubleday, 2007).

39. D. Kapogiannis et al., "Cognitive and Neural Foundations of Religious Belief," *Proceedings of the National Academy of Sciences, Early Edition* 106 (February 3, 2009): 4876–81; see also R. Wright, *The Evolution of God* (Boston: Little, Brown, 2009).

40. Basalla, *Civilized Life in the Universe*.

41. F. S. Collins, *The Language of God* (New York: Free Press, 2006).

CHAPTER 9. PHILOSOPHICAL CONSIDERATIONS

1. J. Cornwell, *Newman's Unquiet Grave* (London: Continuum, 2010).

2. P. Radin, *Primitive Man as Philosopher* (1927; repr., New York: Dover, 1957).

3. *New Columbia Encyclopedia* (New York: Columbia University Press, 1975), s.v. "manichaeans."

4. J. M. Scher, ed., *Theories of the Mind* (New York: Free Press, 1962); see also T. S. Hall, *Ideas of Life and Matter*, vol. 2 (Chicago: University of Chicago Press, 1969).

5. J. Beahrs, personal communication with the author.

6. G. H. Mead. *On Social Psychology* (Chicago: University of Chicago Press, 1956), p. 88.

7. A. Damasio, *Descartes' Error* (1994; repr., London: Vintage, 2005), p. 248.

8. G. Ryle, *The Concept of Mind* (1949; repr., Chicago: University of Chicago Press, 2002).

9. H. Plotkin, *Darwin Machines and the Nature of Knowledge* (Cambridge, MA: Harvard University Press, 1994).

10. R. Shorto, *Descartes' Bones* (New York: Doubleday, 2008).

11. P. Bloom, *Descartes' Baby* (New York: Free Press, 2004); see also M. Shermer, *The Believing Brain* (New York: Times Books, 2011).

12. S. Gaidos, "Going Under," *Science News*, May 21, 2011.

13. J. A. Fodor, *The Mind Doesn't Work That Way: The Scope and Limits of Computational*

Psychology (Cambridge, MA: MIT Press, 2000); see also J. A. Fodor, *RePresentations: Philosophical Essays on the Foundations of Cognitive Science* (Cambridge, MA: MIT Press, 1978).

14. R. Menary, ed., *The Extended Mind* (Cambridge, MA: MIT Press, 2010).

15. R. B. Adams Jr. et al., "Effects of Gaze on Amygdala Sensitivity to Anger and Fear Faces," *Science* 300 (2005): 1536.

16. P. J. Whalen et al., "Human Amygdala Responsivity to Masked Fearful Eye Whites," *Science* 306 (2004): 2061.

17. M. J. Raleigh et al., "Social and Environmental Influences on Blood Serotonin Concentrations in Monkeys," *Archives of General Psychiatry* 41 (1984): 405–10.

18. L. Pessoa, "State of the Art," *Dialogues in Clinical Neuroscience* 12 (2010): 433–48.

19. S. Gaidos, "Cerebral Delights," *Science News*, February 26, 2011).

20. D. G. Myers, "Theories of Emotion," in *Psychology*, 7th ed., ed. D. G. Myers (New York: Worth Publishers, 2004).

21. N. H. Frijda, *The Emotions* (Cambridge: Cambridge University Press, 1986).

22. Damasio, *Descartes' Error*; see also Myers, "Theories of Emotion."

23. D. L. Schacter, *Searching for Memory* (New York: Basic Books, 1996).

24. A. Bechara et al., "Deciding Advantageously before Knowing the Advantageous Strategy," *Science* 275 (1997): 1293–94; see also Damasio, *Descartes' Error*.

25. A. Damasio, *The Feeling of What Happens: Body and Emotion in the Making of Consciousness* (New York: Harcourt, 1999).

26. P. Ekman, "Universals and Cultural Differences in Facial Expressions of Emotion," in *Nebraska Symposium on Motivation 1971*, ed. J. Cole, 19 (1972): 207–83.

27. M. Eid and D. Diener, "Norms for Experiencing Emotions in Different Cultures: Inter- and Intranational Differences," *Journal Personality and Social Psychology* (2001): 36–45.

28. R. J. Davidson et al., "The Privileged Status of Emotion in the Brain," *Proceedings of the National Academy of Sciences* (August 17, 2004): 11915–16.

29. S. Harris et al., "The Neural Correlates of Religious and Nonreligious Belief," *Public Library of Science* (September 30, 2009), http://www.plosone.org/article/info%3 Adoi%2F10.1371%2Fjournal.pone.0007272 (accessed June 5, 2013); see also S. Harris et al., "Functional Neuroimaging of Belief, Disbelief, and Uncertainty," *Annals of Neurology* 63 (2008): 141–47.

30. V. Goel and R. J. Dolan, "Explaining Modulation of Reasoning by Belief," *Cognition* 87 (2003): B11–B21.

31. J. Kluger, "The Biology of Belief," *Time*, February 12, 2009.

32. B. Johnstone and B. A. Glass, "Support for a Neuropsychological Model of Spirituality in Persons with Traumatic Brain Injury," *Zygon* 43 (2008): 861–74.

33. R. Saxe, "Moral Judgments Can Be Altered by Disrupting Specific Brain

Regions," *Medicine & Health/Neuroscience*, March 29, 2010; see also L. Young, "Emotions Key to Judging Others," *Medicine & Health/Neuroscience*, March 24, 2010; J. R. Feierman, "Pedophilia: Its Relationship to Homosexualities and the Roman Catholic Church," *Antonianum* 85 (2010): 451–57.

34. K. R. Ridderinkhof et al., "Alcohol Consumption Impairs Detection of Performance Errors in Mediofrontal Cortex," *Science* 298 (2002): 2209–11.

35. R. R. Griffiths et al., "Psilocybin Can Occasion Mystical-type Experiences Having Substantial and Sustained Personal Meaning and Spiritual Significance," *Psychopharmacology* 187 (2006): 268–83.

36. P. J. Zak, "The Neurobiology of Trust," *Scientific American* (June 2008).

37. J. S. Allen, *The Lives of the Brain* (Cambridge, MA: Harvard University Press, 2009).

38. M. E. Raichie, "The Brain's Dark Energy," *Science* 314 (2006): 1249–50.

39. M. A. Killingworth and D. T. Gilbert, "A Wandering Mind Is an Unhappy Mind," *Science* 330 (2010): 932.

CHAPTER 10. AWARENESS, BELIEF, AND THE PHYSICAL BRAIN

1. R. A. Burton, *On Being Certain* (New York: St. Martin's Griffin, 2008), p. 217.

2. Ibid., p. 218.

3. *Merriam-Webster's Collegiate Dictionary*, 11th ed. (Springfield, MA: Merriam-Webster, 2006), s.v. "awareness."

4. M. Jeannerod, "Consciousness of Action as an Embodied Consciousness," in *Does Consciousness Cause Behavior?* ed. S. Pockett et al. (Cambridge, MA: MIT Press, 2006), pp. 25–38; see also G. Miller, "Feedback from Frontal Cortex May Be a Signature of Consciousness," *Science* 332 (2011): 779.

5. B. Bruya, ed., *Effortless Attention* (Cambridge, MA: MIT Press, 2010); see also T. Bayne, "Phenomenology and the Feeling of Doing: Wegner on the Conscious Will," in Pockett, *Does Consciousness Cause Behavior?*, pp. 169–86.

6. Jeannerod, "Consciousness of Action as an Embodied Consciousness."

7. S. Pockett, "The Neuroscience of Movement," in Pockett, *Does Consciousness Cause Behavior?*, pp. 9–24.

8. Burton, *On Being Certain*; see also Jeannerod, "Conciousness of Action as an Embodied Consciousness."

9. B. Libert, "Time of Conscious Intention to Act in Relation to Onset of Cerebral Activity (Readiness-Potential): The Unconscious Initiation of a Freely Voluntary Act," *Brain* 106 (1983): 1216–28; see also D. M. Eagleman, "The Where and When of Intention," *Science* 303 (2004): 1144–46; M. Desmurget et al., "Movement Intention after Parietal

Cortex Stimulation in Humans," *Science* 324 (2009): 811–13; and P. Haggard, "The Sources of Human Volition," *Science* 324 (2009): 731–33.

10. R. Menary, ed., *The Extended Mind* (Cambridge, MA: MIT Press, 2010).

11. G. D. Stuber et al., "Reward-Predictive Cues Enhance Excitatory Synaptic Strength onto Midbrain Dopamine Neurons," *Science* 321 (2008): 1690–92; see also G. D. Stuber et al., "Excitatory Transmission from the Amygdala to Nucleus Accumbens Facilitates Reward Seeking," *Nature* 475 (2011): 377–80; K. Matsumoto et al., "Neuronal Correlates of Goal-Based Motor Selection in the Prefrontal Cortex," *Science* 301 (2003): 229–32; and S. B. Flagel et al., "A Selective Role for Dopamine in Stimulus-Reward Learning," *Nature* 469 (2011): 53–57.

12. P. Cisek and J. F. Kalaska, "Neural Correlates of Mental Rehearsal in Dorsal Premotor Cortex," *Nature* 431 (2004): 993–96; M. R. Roesch and C. R. Olson, "Neuronal Activity Related to Reward Value and Motivation in Primate Frontal Cortex," *Science* 304 (2004): 307–10; and Eagleman, "Where and When of Intention."

13. L. P. Sugrue et al., "Matching Behavior and the Representation of Value in the Parietal Cortex," *Science* 304 (2004): 1782–86; see also H. Phillips, "The Pleasure Seekers," *New Scientist*, October 11, 2003.

14. M. Shermer, *The Believing Brain* (New York: Times Books, 2011).

15. R. Custers and H. Aarts, "The Unconscious Will: How the Pursuit of Goals Operates outside of Conscious Awareness," *Science* 329 (2010): 47–50; see also P. Berkes et al., "Spontaneous Cortical Activity Reveals Hallmarks of an Optimal Internal Model of the Environment," *Science* 331 (2011): 83–87.

16. M. A. Killingworth, "A Wandering Mind Is an Unhappy Mind," *Science* 330 (2010): 932.

17. K. Smith, "Taking Aim at Free Will," *Nature* 477 (2011): 23–25; see also L. Baumeister et al., "Do You Believe in Free Will?" *PsyBlog*, January 23, 2009, http://www.spring.org.uk/2009/01/do-you-believe-in-free-will.php (accessed April 3, 2010); D. Mobbs, "Free Will Takes Flight: How Our Brains Respond to an Approaching Menace," *Phys.org*, August 23, 2007, http://www.physorg.com/news107098087.html (accessed August 24, 2007); D. Overbye, "Free Will: Now You Have It, Now You Don't," *New York Times*, January 2, 2007; D. Kahneman, *Thinking Fast and Slow* (New York: Farrar, 2011); M. Gazziniga, *Who's in Charge? Free Will and the Science of the Brain* (New York: Ecco, 2011); and D. Eagleman, *Incognito* (New York: Pantheon, 2011).

18. G. Dragoi and S. Tonegawa, "Preplay of Future Place Cell Sequences by Hippocampal Cellular Assemblies," *Nature* 469 (2011): 397–401, see also E. I. Moser and M. B. Moser, "Seeing into the Future," *Nature* 469 (2011): 303–304; and Cisek and Kalaska, "Neural Correlates of Mental Rehearsal in Dorsal Premotor Cortex."

19. S. Gallagher, "Where's the Action? Epiphenomenalism and the Problem of Free Will," in Pockett, *Does Consciousness Cause Behavior?*, pp. 109–24; see also M. Farah, "Area

Responsible for 'Self-Control' Found in the Human Brain," *Phys.org*, August 21, 2007, http://www.physorg.com/news106936688.html (accessed August 23, 2007).

20. Burton, *On Being Certain*.

21. A. Noe, *Action in Perception* (Cambridge, MA: MIT Press, 2004); see also M. A. Palmer, "Beyond Infrastructure," *Nature* 467 (2010): 534–35; and L. Sanders, "Residents of the Brain," *Science News*, July 30, 2011.

22. S. Harris et al., "Functional Neuroimaging of Belief, Disbelief, and Uncertainty," *Annals of Neurology* 63 (2008): 141–47.

23. These choices provide an interpretative quandary.

24. J. L. Vincent et al., "Intrinsic Functional Architecture in the Anaesthetized Monkey Brain," *Nature* 447 (2007): 83–86.

25. Ibid.; see also M. A. Pinsk and S. Kastner, "Unconscious Networking," *Nature* 447 (2007): 46–47.

26. Moser and Moser, "Seeing into the Future."

27. J. Feierman, personal communication with the author; see also Moser and Moser, "Seeing into the Future."

28. W. J. Freeman, "Consciousness, Intentionality, and Causality," in Pockett, *Does Consciousness Cause Behavior?*, pp. 73–108; see also K. Meyer, "Another Remembered Present," *Science* 335 (2012): 415–16; and Eagleman, *Incognito*.

29. L. Floridi, *The Philosophy of Information* (Oxford: Oxford University Press, 2010).

30. C. E. Shannon and W. Weaver, *The Mathematical Theory of Communication* (Urbana: University of Illinois Press, 1949).

CHAPTER 11. THE BIOLOGY OF BELIEF

1. R. D. Alexander, "The Search for a General Theory of Behavior," *Behavioral Sciences* 20 (1975): 77–100.

2. T. E Cerling et al., "Woody Cover and Hominin Environments in the Past 6 Million Years," *Nature* 476 (2011): 51–56.

3. J. Krause et al., "The Complete Mitochondrial DNA Genome of an Unknown Hominin from Southern Siberia," *Nature* 464 (2010): 894–97; see also E. Callaway, "Fossil Genome Reveals Ancestral Link," *Nature* 468 (2010): 1012; D. Reich et al., "Genetic History of an Archaic Hominin Group from Denisova Cave in Siberia," *Nature* 468 (2010): 1053–60; S. A. Parfitt et al., "Early Pleistocene Human Occupation at the Edge of the Boreal Zone in Northwest Europe," *Nature* 466 (2010): 229–32; A. Gibbons, "A New View of the Birth of *Homo sapiens*," *Science* 331 (2011): 392–94; and R. Dennell, "Early *Homo sapiens* in China," *Nature* 468 (2010): 512–13.

4. Gibbons, "New View of the Birth of *Homo sapiens*."

5. Dennell, "Early *Homo sapiens* in China."

6. M. A. Bell et al., *Evolution since Darwin* (Sunderland, MA: Sinauer, 2010).

7. Lin Edwards, "Humans Were Once an Endangered Species," January 21, 2010, *Phys.org*, http://www.physorg.com/news183278038.html (accessed January 19, 2010).

8. H. Harpending, "Are Humans Evolving Faster?" December 6, 2007, *Phys.org*, http://www.physorg.com/news116169889.html (accessed December 10, 2007).

9. J. Shreeve, "Evolutionary Road," *National Geographic* 218 (2010): 34–61.

10. S. P. McPherron et al., "Evidence for Stone-Tool-Assisted Consumption of Animal Tissues before 3.39 Million Years Ago at Dikika, Ethiopia," *Nature* 466 (2010): 857–60; see also D. R. Braun, "Australopithecine Butchers," *Nature* 466 (2010): 828.

11. J. Kass, "Surprisingly Complex Behaviors Appear to Be 'Hard-Wired' in the Primate Brain," March 15, 2005, EurekAlert!, http://www.eurekalert.org/pub_releases/2005-03/vu-scb031505.php (accessed March 16, 2005); see also W. R. Clark and M. Grunstein, *Are We Hardwired? The Role of Genes in Human Behavior* (New York: Oxford University Press, 2000).

12. T. H. Saey, "Genetic Dark Matter," *Science News* (December 18, 2010); see also T. H. Saey, "Scientists Still Making Entries in Human Genetic Encyclopedia," *Science News*, November 6, 2010.

13. J. Z. Tsien, "Brain's Reward Center Also Responds to Bad Experiences," *Medicine & Health/Neurosciences*, February 11, 2011; see also S. B. Flagel et al., "A Selective Role for Dopamine in Stimulus-Reward Learning," *Nature* 469 (2011): 53–57.

14. E. Koechlin et al., "The Architecture of Cognitive Control in the Human Frontal Cortex," *Science* 302 (2003): 1181–85.

15. P. J. Zak, "The Neurobiology of Trust," *Scientific American* (June 2008); see also C. K. W. De Dreu et al., "The Neuropeptide Oxytocin Regulates Parochial Altruism in Intergroup Conflict among Humans," *Science* 328 (2010): 1408–11.

16. T. Gruter and C. C. Carbon, "Escaping Attention," *Science* 328 (2010): 435–36; see also A. W. Woolley et al., "Evidence for a Collective Intelligence Factor in the Performance of Human Groups," *Science* 330 (2010): 686–88.

17. L. Brothers, *Friday's Footprint* (New York: Oxford University Press, 1997); see also A. Dranovsky, "Brain Structure Adapts to Environmental Change," *Neuroscience*, Medical Xpress, June 13, 2011, http://medicalxpress.com/news/2011-06-brain-environmental.html (accessed June 13, 2011).

18. A. Gopnik, *The Philosophical Baby* (London: Bodley Head, 2010).

19. B. Bower, "Tool Finishing Technique Arose before Humans Left Africa," *Science News*, November 20, 2010.

20. S. Dehaene, *Reading in the Brain* (New York: Viking, 2009).

21. D. M. Buss, *The Evolution of Desire* (New York: Basic Books, 1994); see also A. Troisi, "Gender Differences in Vulnerability to Social Stress: A Darwinian Perspective,"

Physiology & Behavior 73 (2001): 443–49; S. Baron-Cohen, *The Essential Difference: Men, Women and the Extreme Male Brain* (London: Allen Lane, 2003); S. Baron-Cohen et al., "Sex Differences in the Brain: Implications for Explaining Autism," *Science* 310 (2005): 819–23; D. C. Funder, "Personality," *Annual Review of Psychology* 52 (February 2001), http://www.annualreviews.org/doi/abs/10.1146/annurev.psych.52.1.197 (accessed June 5, 2013); O. Collignon, "Women Outperform Men When Identifying Emotions," *Medicine & Health/Psychology & Psychiatry*, October 21, 2009; M. Eisenstein, "The First Supper," *Nature* 468 (2010): S8–S9; M. Eisenstein, "Of Beans and Genes," *Nature* 468 (2010): S13–S15; and K. MacDonald, "Personality, Evolution, and Development," in *Evolution and Human Development*, ed. R. Burgess and K. F. MacDonald (Thousand Oaks, CA: Sage Publications, 2004), pp. 2–28.

22. R. Boyd and P. J. Richerson, *The Origin and Evolution of Cultures* (New York: Oxford University Press, 2005); see also P. J. Richerson et al., "Cultural Innovations and Demographic Change," *Human Biology* 81 (2009): 211–35.

23. M. D. Hauser, "The Impossibility of Impossible Cultures," *Nature* 460 (2009): 190–96.

24. Parfitt et al., "Early Pleistocene Human Occupation at the Edge of the Boreal Zone in Northwest Europe," *Nature* 466 (2010): 229–32.

25. F. Salter, *On Genetic Interests* (Frankfurt am Main: Peter Lang, 2003); see also E. Pennisi, "Close Encounters of the Prehistoric Kind," *Science* 328 (2010): 680–83; R. E. Green et al., "A Draft Sequence of the Neandertal Genome," *Science* 328 (2010): 710–22; L. Sanders, "Genes Reveal Mysterious Group of Hominoids as Neandertal Relatives," *Science News*, January 15, 2011; A. Gibbons, "Lucy's Big Brother Reveals New Facets of Her Species," *Science* 328 (2010): 1619; T. Bayne, *The Unity of Consciousness* (Oxford: Oxford University Press, 2010); and E. Pennisi, "Human Evolution: Did Cooked Tubers Spur the Evolution of Big Brains?" *Science* 283 (1999): 2004–2005.

26. R. I. M. Dunbar and S. Shultz, "Evolution in the Social Brain," *Science* 317 (2007): 1344–47; see also C. Efferson et al., "The Coevolution of Cultural Groups and Ingroup Favoritism," *Science* 321 (2008): 1844–49; and M. Brune et al., eds., *The Social Brain* (Chichester, UK: Wiley and Sons, 2003).

27. T. Singer et al., "Empathic Neural Responses Are Modulated by the Perceived Fairness of Others," *Nature* 439 (2006): 466–69; see also R. L. Trivers, "The Evolution of Reciprocal Altruism," *Quarterly Review of Biology* 46 (1971): 35–57; E. Fehr and U. Fischbacher, "The Nature of Human Altruism," *Nature* 425 (2003): 785–91; S. M. H. Hamann, "Positive and Negative Emotional Verbal Stimuli Elicit Activity in the Left Amygdala," *Neuroreport* 13 (2002): 15–19; and M. Balter, "Did Working Memory Spark Creative Culture?" *Science* 328 (2010): 160–63.

28. P. B. deMenocal, "Climate and Human Evolution," *Science* 331 (2011): 540–42; see also Dranovsky, "Brain Structure Adapts to Environmental Change."

29. E. Mayr, *Populations, Species, and Evolution* (Cambridge, MA: Harvard University Press, 1940).

30. R. I. M. Dunbar, "Neocortex Size as a Constraint on Group Size in Primates," *Journal of Human Evolution* 22 (1992): 469–93; see also R. I. M. Dunbar, "Coevolution of Neocortical Size, Group Size, and Language in Humans," *Behavioral and Brain Sciences* 16 (1993): 681–735; R. Dunbar, "Evolution of the Social Brain," *Science* 302 (2003): 1160–61; R. Dunbar, "You've Got to Have (150) Friends," *New York Times*, December 25, 2010; W. Zhou et al., "Discrete Hierarchical Organization of Social Group Sizes," *Proceedings: Biological Sciences* 272 (2005): 439–44; and L. Tiger, *The Pursuit of Pleasure* (Boston: Little, Brown, 1992).

31. G. Miller, "Social Savvy Boosts the Collective Intelligence of Groups," *Science* 330 (2010): 22; see also A. Williams et al., "Evidence for a Collective Intelligence Factor in the Performance of Human Groups," *Science* 330 (2010): 686–88; M. O. Ernst, "Decisions Made Better," *Science* 329 (2010): 1022–23; and B. Chapais, "The Deep Social Structure of Humankind," *Science* 331 (2011): 1276–77.

32. R. Sorabji, *Self* (Oxford: Oxford University Press, 2008).

33. R. C. Malenka and R. Malinow, "Recollection of Lost Memories," *Nature* 469 (2011): 44–45.

34. B. Bower, "Talking Alike Cements Relationships," *Science News*, December 18, 2010.

35. D. Bickerton, *Adam's Tongue* (London: Hill and Wang, 2010).

36. M. Tomasello, *Origins of Human Communication* (Cambridge, MA: MIT Press, 2008); see also W. H. Calvin, *The Brief History of the Mind: From Apes to Intellect and Beyond* (New York: Oxford University Press, 2004); and Bickerton, *Adam's Tongue*.

37. P. Lieberman, *The Biology and Evolution of Language* (Cambridge, MA: Harvard University Press, 1984); M. Balter, "Animal Communication Helps Reveal Roots of Language," *Science* 328 (2010): 969–71.

38. Tomasello, *Origins of Human Communication*.

39. Dehaene, *Reading in the Brain*.

40. Balter, "Animal Communication Helps Reveal Roots of Language."

41. M. Dindo et al., "Observational Learning in Orangutan Cultural Transmission Chains," *Royal Society*, September 15, 2010, http://rsbl.royalsocietypublishing.org/content/early/2010/09/10/rsbl.2010.0637 (accessed September 16, 2010).

42. L. Rendell et al., "Why Copy Others? Insights from the Social Learning Strategies Tournament," *Science* 328 (2010): 208–13; see also E. Pennisi, "Conquering by Copying," *Science* 328 (2010b): 165–67; and J. Cooper, "MRI Scans Show Brain's Response to Actions of Others," *Medicine & Health/Psychology & Psychiatry*, August 11, 2010.

43. J. Feierman, personal communication with the author.

44. J. Panksepp, "Affective Consciousness: Core Emotional Feelings in Animals and Humans," *Consciousness and Cognition* (2005).

45. H. Phillips, "The Pleasure Seekers," *New Scientist*, October 11, 2003; see also R. Layard, "Measuring Subjective Well-Being," *Science* 327 (2010): 534–35.

46. L. P. Sugrue et al., "Matching Behavior and the Representation of Value in the Parietal Cortex," *Science* 304 (2004): 1782–86.

47. M. T. McGuire and A. Troisi, "Physiological Regulation-Deregulation and Psychiatric Disorders," in *Ethology and Sociobiology*, ed. J. Feierman, 8 (1987): 9S–12S; see also T. Canli et al., "Amygdala Responses to Happy Faces as a Function of Extraversion," *Science* 296 (2002): 2191; P. J. Whalen et al., "Human Amygdala Responsivity to Masked Fearful Eye Whites," *Science* 306 (2004): 2061; J. J. Patton et al., "The Primate Amygdala Represents the Positive and Negative Value of Visual Stimuli during Learning," *Nature* 439 (2006): 865–70; L. Tiger and M. McGuire, *God's Brain* (Amherst, NY: Prometheus Books, 2010); and R. J. Dolan, "Emotion, Cognition, and Behavior," *Science* 298 (2002): 1191–94.

48. L. Young, "Emotions Key to Judging Others," *Medicine & Health/Neuroscience*, March 24, 2010.

49. Layard, "Measuring Subjective Well-Being."

50. C. Heyes, "Four Routes of Cognitive Evolution," *Psychological Reports* 110 (2003): 713–27.

51. J. N. Wood et al., "The Perception of Rational, Goal-Directed Action in Non-human Primates," *Science* 317 (2007): 1402–1405; see also L. Palmer and G. Lynch, "A Kantian View of Space," *Science* 328 (2010): 1487–88.

52. P. Bloom, *Descartes' Baby* (New York: Free Press, 2004).

53. S. Gaidos, "More Than a Feeling," *Science News*, August 14, 2010; see also M. V. Flinn et al., "Ecological Dominance, Social Competition, and Coalitionary Arms Race," *Evolution and Human Behavior* 26 (2005): 10–46, http://jayhanson.us/_Biology/Social_Arms_Race.pdf (accessed June 6, 2013).

54. A. Rustichini, "Emotion and Reason in Making Decisions," *Science* 310 (2005): 1624–25.

55. C. Stringer and R. McKie, *African Exodus* (London: Jonathan Cape, 1996).

56. M. Lawler, "A Forgotten Corridor Rediscovered," *Science* 328 (2010): 1092–97.

57. L. L. Cavalli-Sforza et al., *The History and Geography of Human Genes* (Princeton, NJ: Princeton University Press, 1994).

58. B. Bower, "The Ultimate Colonists," *Science News*, July 5, 2003.

59. A. Powell et al., "Late Pleistocene Demography and the Appearance of Modern Human Behavior," *Science* 324 (2009): 1298–1301.

60. R. D. Hernandez et al., "Classic Selective Sweeps Were Rare in Recent Human Evolution," *Science* 331 (2011): 920–24.

61. H. Harpending, "Are Humans Evolving Faster?" December 6, 2007, *Phys.org*, http://www.physorg.com/news116169889.html (accessed December 10, 2007).

CHAPTER 12. ENTER IMAGININGS, BELIEFS, UNCERTAINTY, AND AMBIGUITY

1. J. Feierman, "The Image of God to Whom We Pray: An Evolutionary Psycho‐biological Perspective," *Pensamiento* 67254 (2012): 817–29.

2. T. B. Ward et al., *Creative Thought* (Washington, DC: American Psychiatric Association, 1997); see also P. Harris, *The Work of Imagination* (London: Blackwell, 2000).

3. D. M. Eagleman, "The Where and When of Intention," *Science* 303 (2004): 1144–46; see also P. Cisek and J. F. Kalaska, "Neural Correlates of Mental Rehearsal in Dorsal Premotor Cortex," *Nature* 431 (2004): 993–96; and G. D. Stuber et al., "Reward-Predictive Cues Enhance Excitatory Synaptic Strength onto Midbrain Dopamine Neurons," *Science* 321 (2008): 1690–92.

4. B. Y. Hayden et al., "Fictive Reward Signals in the Anterior Cingulated Cortex," *Science* 324 (2009): 948–50.

5. Stuber, "Reward-Predictive Cues Enhance Excitatory Synaptic Strength."

6. G. Lawton, "Let's Get Personal," *New Scientist*, September 13, 2003.

7. P. Brugger, "Normal Beliefs Linked to Brain Chemistry," *New Scientist*, July 27, 2002.

8. F. Van der Veen, "Blame Serotonin Levels for Being a Crybaby," *Times of India*, September 10, 2010.

9. J. W. Buckholtz et al., "Dopaminergic Network Differences in Human Impulsivity," *Science* 329 (2010): 532; see also J. H. Kelsoe, "A Gene for Impulsivity," *Nature* 468 (2010): 1049–50; and L. Bevilacqua et al., "A Population-Specific HTR2B Stop Condon Predisposes to Severe Impulsivity," *Nature* 468 (2010): 1061–66.

10. L. Pessoa, "Seeing the World in the Same Way," *Science* 303 (2004): 1617–18.

11. D. Lewis-Williams and D. Pearce, *Inside the Neolithic Mind: Consciousness, Cosmos and the Realm of the Gods* (London: Thames and Hudson, 2005).

12. H. B. Linton and R. J. Langs, "Subjective Reactions to Lysergic Acid Diethylamide (LSD-25)," *Archives of General Psychiatry* 6 (1962): 352–58; see also M. M. Katz et al., "Characterizing the Psychological State Produced by LSD," *Abnormal Psychology* 73 (1968): 1–14.

13. B. E. Depue et al., "Prefrontal Regions Orchestrate Suppression of Emotional Memories via a Two-Phase Process," *Science* 317 (2007): 215–19; see also M. P. Walker et al., "Dissociable Stages of Human Memory Consolidation and Reconstruction," *Nature* 425 (2003): 616–20; and J. L. C. Lee et al., "Independent Cellular Processes for Hippocampal Memory Consolidation and Reconsolidation," *Science* 304 (2004): 839–43.

14. M. Balter, "Did Working Memory Spark Creative Culture?" *Science* 328 (2010): 160–63.

15. P. Ricoeur, *Memory, History, Forgetting* (Chicago: University of Chicago Press, 2006).

16. L. Backman et al., "Effects of Working-Memory Training on Striatal Dopamine Release," *Science* 333 (2011): 718.

17. M. Ohbayashi et al., "Conversion of Working Memory to Motor Sequence in the Monkey Premotor Cortex," *Science* 301 (2010): 233–36.

18. J. Kagan, *Surprise, Uncertainty, and Mental Structures* (Cambridge, MA: Harvard University Press, 2002).

19. C. A. Anderson, "Belief Perseverance," in *Encyclopedia of Social Psychology*, ed. R. F. Baumeister and K. D. Vohs (Thousand Oaks, CA: Sage, 2007), pp. 109–10.

20. M. R. Roesch and C. R. Olson, "Neuronal Activity Related to Reward Value and Motivation in Primate Frontal Cortex," *Science* 304 (2004): 307–10; see also Eagleman, "Where and When of Intention"; and Cisek and Kalaska, "Neural Correlates of Mental Rehearsal in Dorsal Premotor Cortex."

21. J. G. Frazer, *The New Golden Bough* (1890; repr., New York: Criterion Books, 1959); see also D. Park, *The Grand Contraption: The World of Myth, Number and Chance* (Princeton, NJ: Princeton University Press, 2008).

22. L. Tiger and M. McGuire, *God's Brain* (Amherst, NY: Prometheus Books, 2010).

23. Kagan, *Surprise, Uncertainty, and Mental Structures.*

24. J. D. Cohen and G. Aston-Jones, "Decision amid Uncertainty," *Nature* 436 (2005): 472; see also E. J. Hermans et al., "Stress-Related Noradrenergic Activity Prompts Large-Scale Neural Network Reconfiguration," *Science* 334 (2011): 1151–53; and Tiger and McGuire, *God's Brain.*

CHAPTER 13. THEORY OF MIND, MIRRORING, AND ATTRIBUTION

1. L. B. Steadman et al., "Toward a Testable Definition of Religious Behavior," in *The Biology of Religious Behavior*, ed. J. Feierman (Santa Barbara, CA: Praeger, 2009), pp. 20–35.

2. J. Beahrs, personal communication with the author.

3. R. Dunbar, "Evolution of the Social Brain," *Science* 302 (2003): 1160–61; see also G. Miller, "Reflecting on Another's Mind," *Science* 308 (2005): 945–47.

4. M. Shermer, *The Believing Brain* (New York: Times Books, 2011), p. 3.

5. D. D. Hotto, "Folk Psychological Narratives," (Cambridge, MA: MIT Press, 2008).

6. S. Baron-Cohen, "Precursors to a Theory of Mind: Understanding Attention in Others," in *Natural Theories of Mind: Evolution, Development and Simulation of Everyday Mindreading*, ed. A. Whiten (Oxford: Basil Blackwell, 1991), pp. 233–51; see also A. N. Meltzoff, "Imitation as a Mechanism of Social Cognition: Origins of Empathy, Theory of Mind, and the Representation of Action," *Handbook of Childhood Cognitive Development*, ed. U. Goswami (Oxford: Blackwell Publishers, 2002), p. 6–25.

7. Meltzoff, "Imitation as a Mechanism of Social Cognition."

8. F. C. Keil, "Science Starts Early," *Science* 331 (2011): 1022–23; see also A. Meltzoff, "Understanding the Intentions of Others: Re-enactment of Intended Acts by 18-Month-Old Children," *Developmental Psychology* 31 (1995): 838–50; J. Perner et al., "Identity: Key to Children's Understanding of Belief," *Science* 333 (2011): 474–77; and C. Zimmer, "How the Mind Reads Other Minds," *Science* 309 (2003): 1079–80.

9. R. A. Spitz, "Hospitalism—An Inquiry into the Genesis of Psychiatric Conditions in Early Childhood," *Psychoanalytic Study of the Child* 1 (1945): 53–74.

10. H. Harlow, "Maternal Behavior of Rhesus Monkeys Deprived of Mothering and Peer Associations in Infancy," *Proceedings of the American Philosophical Society* 110 (1959): 421–32.

11. A. Leslie and L. Thaiss, "Domain Specificity in Conceptual Development," *Cognition* 43 (1992): 225–51; see also S. Baron-Cohen et al., "Does the Autistic Child Have a 'Theory of Mind'?" *Cognition* 21 (1985): 37–46; and B. Hare et al., "Do Chimpanzees Know What Conspecifics Know and Do Not Know?" *Animal Behavior* 61 (2001): 139–51.

12. D. G. Premack and C. Woodruff, "Does the Chimpanzee Have a Theory of Mind?" *Behavioral and Brain Sciences* 1 (1978): 515–26; see also R. Dunbar, "Can You Guess What I'm Thinking?" *New Scientist*, June 12, 2004; and J. N. Wood et al., "The Perception of Rational, Goal-Directed Action in Nonhuman Primates," *Science* 317 (2007): 1402–1405.

13. G. Miller, "Probing the Social Brain," *Science* 312 (2006): 838–39.

14. E. Anderson et al., "The Visual Impact of Gossip," *Science* 332 (2011): 1446–48.

15. S. Pinker, *The Blank Slate: The Modern Denial of Human Nature* (London: Allen Lane, 2002).

16. A. Gopnik, *The Philosophical Baby* (London: Bodley Head, 2010).

17. G. Rizzolatti and C. Sinigaglia, *Reflecting on the Mind* (Oxford: Oxford University Press, 2006); see also L. Sanders, "Mirror System Gets an Assist," *Science News*, August 13, 2011; and L. Sanders, "Brain's Mirror System Does the Robot," *Science News*, May 7, 2011.

18. P. S. Churchland, *Braintrust* (Princeton: Princeton University Press, 2011).

19. K. Nakahara and Y. Miyashita, "Understanding Intentions: Through the Looking Glass," *Science* 308 (2005): 644–45; see also L. Fogassi et al., "Parietal Lobe: From Action Organization to Intention Understanding," *Science* 308 (2005): 662–66; and Rizzolatti and Sinigaglia, *Reflecting on the Mind*.

20. T. Singer et al., "Empathy for Pain Involves the Affective but Not the Sensory Components of Pain," *Science* 303 (2004): 1157–62; see also Sanders, "Brain's Mirror System Does the Robot."

21. N. I. Eisenberger et al., "Does Rejection Hurt? An fMRI Study of Social Exclusion," *Science* 302 (2003): 290–92; see also J. Panksepp, "Feeling the Pain of Social Loss," *Science* 302 (2003): 237–39; G. MacDonald and M. R. Leary, "Why Does Social Exclusion Hurt? The Relationship between Social and Physical Pain," *Psychological Bulletin* 131 (2005): 202–23;

and Singer, "Empathy for Pain Involves the Affective but Not the Sensory Components of Pain."

22. Miller, "Reflecting on Another's Mind."

23. S. A. J. Birch and P. Bloom, "The Curse of Knowledge in Reasoning about False Beliefs," *Psychological Science* 18 (June 2007), http://psychologicalscience.org/ (accessed April 9, 2010).

24. Miller, "Reflecting on Another's Mind."

25. L. Festinger, *A Theory of Cognitive Dissonance* (Evanston, IL: Row, Peterson, 1957).

26. T. Tremlin, *Minds and Gods: The Cognitive Foundations of Religion* (Oxford: Oxford University Press, 2006); see also E. Pronin, "How We See Ourselves and How We See Others," Science (2008): 1177–80; and D. R., "The Functions of Attributions," *Social Psychology Quarterly* 43 (1980): 184–89.

27. Shermer, *Believing Brain*.

28. Tremlin, *Minds and Gods*.

29. Pronin, "How We See Ourselves and How We See Others."

30. Forsyth, "Functions of Attributions."

31. M. Walker, "Folk Medicine Poses Global Threat to Primate Species," Earth News, BBC, http://news.bbc.co.uk/earth/hi/earth_news/newsid_8589000/8589551.stm (accessed June 8, 2013).

32. A. Powell et al., "Late Pleistocene Demography and the Appearance of Modern Human Behavior," *Science* 324 (2009): 1298–1301.

CHAPTER 14. STORIES AND MODELS

1. D. Bickerton, *Adam's Tongue* (London: Hill and Wang, 2010).

2. J. Price, "The Adaptiveness of Changing Religious Belief Systems," in *The Biology of Religious Behavior*, ed. J. Feierman (Santa Barbara, CA: Praeger, 2009), pp. 175–89.

3. J. Beahrs, personal communication with the author.

4. G. Currie, *Narratives and Narrators* (Oxford: Oxford University Press, 2010).

5. J. D. Karpicke and J. R. Blunt, "Retrieval Practice Produces More Learning Than Elaborative Studying with Concept Mapping," *Science* 331 (2011): 772–75.

6. B. J. Richmond et al., "Predicting Future Rewards," *Science* 301 (2003): 179–80; see also K. Matsumoto et al., "Neuronal Correlates of Goal-Based Motor Selection in the Prefrontal Cortex," *Science* 301 (2003): 229–32.

7. J. Josipovici, *What Ever Happened to Modernism?* (New Haven, CT: Yale University Press, 2010).

8. K. D. Harrison, *When Languages Die* (New York: Oxford University Press, 2008).

9. V. Stauffer, *The Bavarian Illuminati in America* (1918; repr., Minecola, NY: Dover, 2006).

10. M. Jolly, *Faces of the Living Dead* (West New York, NJ: Mark Batty, 2006).

11. N. Abe et al., "Deceiving Others: Distinct Neural Responses of the Prefrontal Cortex and Amygdala in Simple Fabrication and Deception with Social Interactions," *Journal of Cognitive Neuroscience* 19 (2007): 287–95.

12. R. C. Malenka and R. Malinow, "Recollection of Lost Memories," *Nature* 469 (2011): 44–45.

13. J. Feierman, personal communication with the author.

14. M. Saunders, *Self Impression* (Oxford: Oxford University Press, 2010).

15. Ibid.; see also G. Levin, "A Dive into the Deep Self," *Times Literary Supplement,* April 9, 2010; and J. L. Borges, *On Argentina* (New York: Penguin Classic, 2010).

16. M. M. Hurley et al., *Inside Jokes: Using Humor to Reverse-Engineer the Mind* (Cambridge, MA: MIT Press, 2011).

17. M. de Villiers and S. Hirtle, *Timbuktu* (New York: Walker, 2007).

18. M. D. Jackson, *Social and Economic Networks* (Princeton, NJ: Princeton University Press, 2008).

19. P. Bloom, *Descartes' Baby* (New York: Free Press, 2004).

20. H. Gweon and L. Schulz, "16-Month-Olds Rationally Infer Causes of Failed Actions," *Science* 332 (2011): 1524.

21. B. Bower, "Kids Perceive Ownership Principles," *Science News,* June 18, 2011.

22. B. Bower, "Geometry Comes Naturally to the Unschooled Mind," *Science News,* June 18, 2011.

23. F. C. Keil, "Science Starts Early," *Science* 331 (2011): 1022–23; see also S. Pinker, *The Blank Slate: The Modern Denial of Human Nature* (London: Allen Lane, 2002).

24. M. R. Roesch and C. R. Olson, "Neuronal Activity Related to Reward Value and Motivation in Primate Frontal Cortex," *Science* 304 (2004): 307–10; see also G. D. Stuber et al., "Reward-Predictive Cues Enhance Excitatory Synaptic Strength onto Midbrain Dopamine Neurons," *Science* 321 (2008): 1690–92; Richmond, "Predicting Future Rewards"; and Matsumoto, "Neuronal Correlates of Goal-Based Motor Selection in the Prefrontal Cortex."

25. B. R. J. Jansen et al., "Rule Transition on the Balance Scale Task: A Case Study of Belief Change," *Synthese* 155 (March 2007), http://springerlink.com/content/m24410u27322557q/ (accessed April 3, 2010).

26. P. Berkes et al., "Spontaneous Cortical Activity Reveals Hallmarks of an Optimal Internal Model of the Environment," *Science* 331 (2011): 83–87.

27. M. Ohbayashi et al., "Conversion of Working Memory to Motor Sequence in the Monkey Premotor Cortex," *Science* 301 (2010): 233–36.

28. A. Pickering, *The Cybernetic Brain* (Chicago: University of Chicago Press, 2010).

CHAPTER 15. TRIGGERING

1. J. Feierman, personal communication with the author; see also N. Tinbergen, *The Study of Instinct* (London: Oxford University Press, 1951).

2. I. Eibl-Eibesfeldt, *Human Ethology* (New York: Aldine de Gruyter, 1989).

3. Ibid.

4. J. Feierman, "The Image of God to Whom We Pray: An Evolutionary Psychobiological Perspective," *Pensamiento* 67254 (2012): 817–29.

5. M. A. Pyc and K. A. Rawson, "Why Testing Improves Memory: Mediator Effectiveness Hypothesis," *Science* 330 (2010): 335; see also J. D. Karpicke and J. R. Blunt, "Retrieval Practice Produces More Learning Than Elaborative Studying with Concept Mapping," *Science* 331 (2011): 772–75.

6. G. Xue et al., "Greater Neural Pattern Similarity across Repetitions Is Associated with Better Memory," *Science* 330 (2010): 97–101.

7. G. Dragoi and S. Tonegawa, "Preplay of Future Place Cell Sequences by Hippocampal Cellular Assemblies," *Nature* 469 (2011): 397–401.

8. S. Gaidos, "More Than a Feeling," *Science News*, August 14, 2010.

9. P. K. Kuhl, "Who's Talking?" *Science* 333 (2011): 1311.

10. C. R. Sunstein, *Going to Extremes: How Like Minds Unite and Divide* (Oxford: Oxford University Press, 2009).

11. A. M. Kovacs et al., "The Social Sense: Susceptibility to Others' Beliefs in Human Infants and Adults," *Science* 330 (2010): 830–34.

12. K. C. Bickart et al., "Amygdala Volume and Social Network Size in Humans," *Nature Neuroscience* 14 (2011): 163–64.

CHAPTER 16. INTRANSIGENT BELIEFS AND BELIEF-DISCONFIRMATION FAILURE

1. J. Bruner, *On Knowing* (Cambridge, MA: Harvard University Press, 1962).

2. C. A. Anderson, "Belief Perseverance," in *Encyclopedia of Social Psychology*, ed. R. F. Baumeister and K. D. Vohs (Thousand Oaks, CA: Sage, 2007), pp. 109–10.

3. R. Dawkins, *The God Delusion* (New York: Bantam Books, 2006).

4. M. Edelson et al., "Following the Crowd: Brain Substrates of Long-Term Memory Conformity," *Science* 333 (2011): 108–11.

5. P. Harrigan, "Dionysus and Kataragama: Parallel Mystery Cults," *The Journal of the Institute of Asian Studies*, http://kataragama.org/research/dionysus.htm (accessed June 10, 2013).

6. E. Haney, "Cult Activity in the '90s," Infoplease, http://www.infoplease.com/spot/jonestown3.html (accessed March 8, 2005).

7. Ibid.

8. J. R. Hall, *Gone from the Promised Land: Jonestown in American Cultural History* (New Brunswick, NJ: Transaction Publishers, 1987); see also D. Layton, *Seductive Poison* (New York: Anchor, 1999); and C. Wessinger, *How the Millennium Comes Violently: From Jonestown to Heaven's Gate* (Syracuse, NY: Seven Bridges Press, 2000).

9. S. Yousafzai and R. Moreau, "The Taliban in Their Own Words," *Newsweek*, November 5, 2009; see also R. Ibrahim, "An Analysis of Al-Qa'ida's Worldview," *Middle East Quarterly* (January 6, 2009); Y. H. A. S. Zuhur, "Islamic Rulings on Warfare," *Strategic Studies Institute* (October 2004); B. Thornton, "Ten Years of Lessons Unlearned," *Frontpage Magazine*, September 19, 2011, http://frontpagemag.com/2011/bruce-thornton/ten -years-of-lessons-unlearned/ (accessed June 10, 2013); D. MacEoin, "Suicide Bombing as Worship," *Middle East Quarterly* (Fall 2009); and "Shariah: The Threat to America," (Washington, DC: Center for Security Policy, 2010).

10. A. Rashid, "The Taliban: Exporting Extremism," *Foreign Affairs* (November/ December, 1999).

11. Ibrahim, "Analysis of Al-Qa'ida's Worldview"; see also Thornton, "Ten Years of Lessons Unlearned."

12. G. Friedman, "Germany and the Failure of Multiculturalism," *Stratfor Global Intelligence*, October 19, 2010.

13. J. Couzin-Frankel, "Court to Weigh University's Decision Not to Hire Astronomer," *Science* 330 (2010): 1731; see also B. Waltke, "Evangelical Scholar Expelled over Evolution," National Center for Science Education, April 12, 2012, http://ncse.com/ news/2010/04/evangelical-scholar-expelled-over-evolution-005432 (accessed March 14, 2010).

14. R. Petty, "When People Feel Powerful, They Ignore New Opinions," February 14, 2008, *Phys.org*, http://www.physorg.com/news122212997.html (accessed April 15, 2010).

15. B. Ehrenreich, *Smile or Die* (New York: Granta Books, 2010).

16. J. Beahrs, personal communication with the author.

17. G. Martin and J. Pear, *Behavior Modification: What It Is and How to Do It*, 8th ed. (Upper Saddle River, NJ: Pearson Prentice Hall, 2007).

18. L. D. Strieker, *Brainwashing, Cults, and Deprogramming in the '80s* (New York: Doubleday, 1984).

19. L. Santos, "Human Prejudice Has Ancient Evolutionary Roots," *Other Sciences/ Social Sciences*, March 17, 2011.

20. C. Freeman, *The Closing of the Western Mind* (New York: Vintage, 2005); see also I. F. McNeely and L. Wolverton, *Reinventing Knowledge* (New York: Norton, 2008).

21. D. Gilbert, "Buried by Bad Decisions," *Nature* 474 (2011): 275–77.

22. D. R. Oxley et al., "Political Attitudes Vary with Physiological Traits," *Science* 321 (2008): 1667–70.

23. S. A. J. Birch and P. Bloom, "The Curse of Knowledge in Reasoning about False Beliefs," Association for Psychological Science, http://psychologicalscience.org/ (accessed April 9, 2010).

24. D. Fox, "Brain Buzz," *Nature* 472 (2011): 156–58.

CHAPTER 17. AND WHY?

1. C. Taylor, *A Secular Age* (Cambridge, MA: Harvard University Press, 2007).

2. J. Groopman, "Health Care: Who Knows Best?" *New York Review*, February 11, 2010.

3. J. Gleick, "Information Everything," *Discover*, July 8, 2011; see also E. Pennisi, "Will Computers Crash Genomics?" *Science* 331 (2011): 666–68.

4. P. Maass, "How the Media Inflated a Minor Moment in a Long War," *New Yorker*, January 10, 2011; see also N. Carr, *The Shallows* (New York: Atlantic, 2011).

5. J. H. Billington, "The Unrealized Potential of the Internet," *Bohemian Club Library Notes* 134 (Winter 2005); see also Carr, *Shallows*.

6. Sparrow, 2011. Source lost.

7. J. Rosen, "The End of Forgetting," *New York Times Magazine*, July 25, 2010.

8. N. Bilton, *I Live for the Future & Here's How It Works: Why Your World, Work, and Brain Are Being Creatively Disrupted* (New York: Crown, 2010).

9. L. Tiger, *Men in Groups* (New York: Random House, 1969); see also T. Sharot et al., "Neural Mechanisms Mediating Optimism Bias," *Nature* 450 (2007): 102–105.

10. P. Webb. "Science Education and Literacy: Imperatives for the Developed and Developing World," *Science* 328 (2010): 448–50.

11. A. Stirling, "Keep It Complex," *Nature* 468 (2010): 1029–31; see also R. Schenkel, "The Challenge of Feeding Scientific Advice into Policy-Making," *Science* 330 (2010): 1749–51.

12. S. Halpern, "Mind Control & the Internet," *New York Review*, June 23, 2011; and E. Pariser, *The Filter Bubble: What the Internet Is Hiding from You* (New York: Penguin, 2011).

13. Halpern, "Mind Control & the Internet."

14. D. Fox, "Brain Buzz," *Nature* 472 (2011): 156–58.

15. L. Tiger and M. McGuire, *God's Brain* (Amherst, NY: Prometheus Books, 2010).

16. K. Ohlson, "The End of Morality," *Discover*, July 8, 2010.

17. E. Dias-Ferreira et al., "Chronic Stress Causes Frontostriatal Reorganization and Affects Decision Making," *Science* 325 (2009): 621–25.

18. R. Dawkins, *The God Delusion* (New York: Bantam Books, 2006); see also S. Harris, *The Moral Landscape* (New York: Free Press, 2010).

19. J. T. Fraser, *Time, Conflict, and Human Values* (Urbana: University of Illinois Press, 1999).

20. S. Ozawa et al., "Coseismic and Postseismic Slip of the 2011 Magnitude-9 Tohoku-Oki Earthquake," *Nature* 475 (2011): 373–76.

21. Webb, "Science Education and Literacy"; see also M. A. Korb and U. Thakkar, "Facilitating Scientific Investigations and Training Data Scientists," *Science* 333 (2011): 534–35.

22. E. Schrecker, *The Lost Soul of Higher Education: Corporatization, the Assault on Academic Freedom, and the End of the American University* (New York: New Press, 2010); see also D. Ravich, *The Death and Life of the Great American School System* (New York: Basic Books, 2010); and H. Radder, *The Commodification of Academic Research* (Pittsburgh: University of Pittsburgh Press, 2010).

23. K. Thomas, "What Are Universities For?" *Times Literary Supplement*, May 7, 2010; see also M. S. Nussbaum, "Skills for Life," *Times Literary Supplement*, April 30, 2010.

24. J. Rasenberger, *America, 1908* (New York: Scribner, 2007).

25. M. Hoffmann et al., "The Impact of Conservation on the Status of the World's Vertebrates," *Science* 330 (2010): 1503–1509; see also H. M. Pereira et al., "Scenarios for Global Biodiversity in the 21st Century," *Science* 330 (2010): 1496–1501; C. J. Vorosmarry et al., "Global Threats to Human Water Security and River Biodiversity," *Nature* 467 (2010): 555–61; M. A. Palmer, "Beyond Infrastructure," *Nature* 467 (2010): 534–35; J. M. Drake and B. D. Griffen, "Early Warning Signals of Extinction in Deteriorating Environments," *Nature* 467 (2010): 456–59; and A. D. Barnosky et al., "Has the Earth's Sixth Mass Extinction Already Arrived? *Nature* 471 (2011): 51–57.

CHAPTER 18. WHAT TO DO?

1. J. Mervis, "Report Alters Definition of What Students Should Learn," *Science* 333 (2011): 510; see also M. A. Korb and U. Thakkar, "Facilitating Scientific Investigations and Training Data Scientists," *Science* 333 (2011): 534–35; P. Hines et al., "Laying the Foundation for Lifetime Learning," *Science* 333 (2011): 951–83; and M. Ridley, *The Rational Optimist* (New York: Harper, 2010).

2. S. Harris, *The Moral Landscape* (New York: Free Press, 2010); see also M. Shermer, *The Believing Brain* (New York: Times Books, 2011); and P. Hefner, "It's All about Transforming Minds," *Zygon* 40 (2005): 263–66.

INDEX